MODULES
A PRIMER OF STRUCTURE THEOREMS

TOM HEAD

University of Alaska at Fairbanks
University of Texas at El Paso

BROOKS/COLE PUBLISHING COMPANY
Monterey, California
A Division of Wadsworth Publishing Company, Inc.

To Eileen,
who brings me good luck

ISBN: 0-8185-0109-X
L.C. Catalog Card No.: 73-88319
Printed in the United States of America

1 2 3 4 5 6 7 8 9 10—78 77 76 75 74

PREFACE

This book is designed for use as a text at either the upper-division or graduate level. It is designed for students who have had experience with linear algebra and a solid semester of modern algebra. No previous acquaintance with modules is assumed.

To the Student. The approach I have taken to module structure crystallized in the early 1960s. It was made possible by three papers, Eckmann and Schopf [1953], Matlis [1958], and Kaplansky [1958], and the initial sections of a book of mammoth scope, Cartan and Eilenberg's *Homological Algebra* [1956]. Further results on module structure appear regularly, but I believe that the material presented in this book constitutes a firm central core for the theory of module structure.

Module theory is closely related to other branches of algebra such as homological algebra, commutative and noncommutative rings, group representations, and infinite Abelian groups. Some of these relationships are discussed in Appendix 1 of Chapter 0 and in the notes following Chapters 3, 4, 5, 13, and 14.

In my treatment of module theory, I do not use I to denote the integers. I use I to denote an index set (which may be uncountable). I use $A \setminus B$ to denote $\{a \in A | a \notin B\}$. I advise you to read the exercises whether you have time to work them or not. The heart of the book is divided into three parts. Each time you finish a part, reread the whole part in one sitting, skipping proofs whenever they break the continuity of your thought. Items in the bibliography are referred to by author's name followed by the date of publication in brackets. You are invited to examine these references, but they may be ignored entirely. Don't be discouraged by the fact that a few items in the bibliography demand more background than this text provides.

To the Instructor. Courses consisting of various amounts of material can be taught meaningfully from this book. Chapters 1 through 5 should be considered as a basic unit. The remainder of the book can be divided into the following units of material: (1) Chapter 6 (in detail) plus lectures describing the fundamental results of Chapters 7 and 9; (2)

Chapters 6 through 9; (3) Chapter 11 (in detail) plus Chapters 12 and 13, omitting the proofs of Theorem 8 and/or Proposition 8; (4) Chapters 11 through 13; and (5) Chapter 10. Each of these groupings depends on the coverage of Chapters 1 through 5. In addition, Chapter 10 depends on the coverage of Chapters 6 through 9. A simple procedure is to cover Chapters 1 through 5 and then plan the rest of the course on the basis of the amount of time remaining. For students with minimal background in algebra, Chapters 1 through 5 may be sufficient when Chapter 0 and the exercises are treated thoroughly. A possible conclusion for such a course would be a presentation of Artin [1950]. In any case, the basic unit plus units (1) and (3) would provide adequate challenge for an undergraduate semester. If desired, Chapter 10 can be covered by using a conventional proof of Theorem 5. In a graduate semester it is possible to cover the entire book.

I suggest progressing through Chapter 0 while presenting Chapter 1 and Part One. The exercises in Chapter 0 are placed with the sections they accompany to facilitate independent use of these sections. Few exercises are given in Chapter 0 because an understanding of the formalisms of Chapter 0 is best gained through using the formalisms in later chapters. I suggest spending as much time as students require for a solid grasp of Part One. Once students have worked enough problems in Part One, it is safe to proceed with less emphasis on the exercises.

In three sections of this book I have used infinite ordinal numbers: Chapter 4, §3; Chapter 7, §3; and Chapter 12, §2. The presence of ordinals in these sections is indicated by the abbreviation "Ord." appearing in the section titles. I have not assumed that all or even the majority of readers will have previous familiarity with infinite ordinals. I have not included an exposition of ordinals; instead, I have written the text so that it is consistent with hierarchy of possible teaching strategies regarding ordinals:

1. The proofs in the Ord. sections can be omitted without disturbing the continuity or integrity of the course.
2. The proofs in the first two Ord. sections can be read by the student with virtually complete comprehension even without prior acquaintance with infinite ordinals. (Compare § 2 and § 3 of Chapter 4.) Such reading can motivate careful study of ordinals in a later course.
3. The use of ordinals in algebra can be taught after a lecture or reading assignment covering only the most elementary properties of the ordinal numbers.

My attitude about the use of equivalents of the axiom of choice in teaching from this book is indicated in Chapter 0, § 11.

Although this text was designed for use in the standard senior and first-year graduate courses in algebra, I also encourage you to try it for less usual courses. For example, it may be used (1) in series with texts treating linear algebra and/or group representations by module theoretic means to provide a year course in algebra integrated around the module concept, (2) as the initial increment in series with introductory texts in homological algebra, (3) in parallel with texts in homological algebra, and (4) in student and faculty seminars.

Acknowledgements. This book is a product of work done at Iowa State University, The University of Alaska at Fairbanks, New Mexico State University, and The University of Texas at El Paso. I am grateful for the support given me by these universities and for the encouragement I have received from colleagues at these institutions. I thank Joseph Hershenov of Queens College for arranging for my use of office space and library during a summer of work on this book.

The work of Richard Mitchell and Roger Mitchell of the University of Texas, Arlington, and Robert Warfield of the University of Washington in reviewing the manuscripts has been invaluable. I thank these reviewers and the series editor, Robert Wisner of New Mexico State University, for their support and encouragement. In the labor of transferring my thoughts from handwriting to print, I have been blessed with two excellent typists, Tess Pearce and Gloria Valdez, and a helpful editorial staff at Brooks/Cole, including Ginny Decker. For their advice on special points in the text, I thank Delmar Boyer, Eileen Head, William Leahey, and Edgar Rutter.

Tom Head

CONTENTS

INTRODUCTION

CHAPTER 0

MODULES, MAPS, AND DIRECT SUMS

§ 1 MODULES

Each ring R that we consider will be assumed to contain a multiplicative identity element, which will be denoted 1. We will therefore regard the possession of such an identity as one of the defining conditions of the ring concept. We will also assume $1 \neq 0$. Now let R be a ring.

DEFINITION. A *left* R-*module* consists of an Abelian group A together with a function $m : R \times A \to A$ that satisfies the following four conditions:

(1) $m(r, (a + a')) = m(r, a) + m(r, a')$,
(2) $m((r + r'), a) = m(r, a) + m(r', a)$,
(3) $m(rr', a) = m(r, m(r', a))$, and
(4) $m(1, a) = a$,

where r and r' are arbitrary elements of R and a and a' are arbitrary elements of A.

The element $m(r, a)$ of A is conventionally denoted $r \cdot a$ or ra, and m is said to provide a scalar operation of the ring R on the Abelian group A. When the function m is replaced by the juxtaposition notation, the four conditions above become:

(1) $r(a + a') = ra + ra'$,
(2) $(r + r')a = ra + r'a$,
(3) $(rr')a = r(r'a)$, and
(4) $1a = a$.

When it is clear which ring R is under discussion, the expression "left R-module" will be abbreviated to "left module." In Appendix 2 of this chapter we will discuss *right* modules and their relationship with *left* modules, but within the body of this book all modules considered will be left modules. Consequently, "left module" is abbreviated to "module."

A purist will observe that a left R-module is an ordered pair (A, m) where the first member, A, is an Abelian group and the second member, m, is a function. In referring to the module (A, m), it will always be sufficiently clear to mention only the Abelian group A, deleting reference to the function m. Finally, the definition should not be construed to represent any special preference for the letter A to represent a module. Thus, I will continually begin discussions with such (hopefully) unambiguous statements as "Let M and N be R-modules." Packed into such a statement are the assumptions that R is a ring with multiplicative identity, that M and N are Abelian groups, and that there are functions $m : R \times M \to M$ and $m' : R \times N \to N$ that satisfy conditions 1 through 4 of the definition stated above. We are ready for some examples of modules.

Modules over a *field* R are also called *vector spaces over* R. Elementary courses in linear algebra deal with vector spaces over the field R of real numbers and consequently supply an immediate collection of examples of R-modules.

Each Abelian group A may be regarded as a module over the ring Z of integers in precisely one way: for z in Z and a in A, define za by

$$za = \begin{cases} \text{if } z > 0: & \text{the sum of } z \text{ copies of } a, \\ \text{if } z = 0: & 0, \\ \text{if } z < 0: & \text{the sum of } |z| \text{ copies of } -a. \end{cases}$$

It is elementary to verify that the four conditions specified in the definition of a module are satisfied by this operation and that this is the only operation of Z on A that is compatible with these conditions (see Exercise 1b). Because each Abelian group allows precisely one Z-module structure (and because of some additional elementary facts to be pointed out in this chapter), the theory of Abelian groups is indistinguishable from the theory of Z-modules and is thereby subsumed by module theory.

Each ring R provides an example of an R-module in the following sense: for the Abelian group, use R with its ring addition. For the scalar multiplication of the ring R on the Abelian group R, use the ring multiplication. Each of the four conditions in the definition of an R-module follows from one of the usual defining conditions in the definition of a ring (see Exercise 2). From the point of view developed in this book, the ring R regarded in this way as a module over itself is the starting point for the structure theory of R-modules. Chapter 1 is devoted entirely to an examination of R as a module over itself, and the remainder of the book is an investigation of how further R-modules can be developed from R itself by means of constructions described in the present chapter and one additional construction introduced in Chapter 7.

Why are modules worthy of study? As you should expect, it is not easy to answer this question at this point. Appendix 1 of the present chapter is devoted to this question and may be consulted at any time, although it probably cannot be fully understood before the completion of Part One.

EXERCISES

1. a. Show that the operation of the ring Z of integers on the Abelian group A (as described on page 3) provides a Z-module structure for A.
   b. Show that this is the only Z-module structure that A admits.
2. Verify that the four conditions listed in the definition of a module are satisfied by the operation of the ring R on the additive Abelian group R that is provided by the ring multiplication (as discussed above).

§ 2 SUBMODULES AND QUOTIENT MODULES

Let A be a module over a ring R. A subset S of A is a submodule of A if S is a subgroup of A and rs is in S whenever r is in R and s is in S. If R is a field, then the submodules of each R-module (that is, vector space over R) are precisely the familiar subspaces.

Let A be an arbitrary Abelian group. Regard A as a module over the ring Z of integers, as described in the previous section. Observe that from the definition of the operation of Z on A it follows that every subgroup of A is necessarily a submodule. Thus, for Abelian groups (that is,

Z-modules), the subgroups and submodules are precisely the same subsets. Without this fact we could not fully identify the theory of Abelian groups with the theory of Z-modules.

Now let R be an arbitrary ring and let B be a submodule of an R-module A. Since B is a subgroup of the Abelian group A, the quotient group A/B is available. How does the scalar operation of R on A provide a scalar operation of R on A/B? Let C be an element of A/B (that is, a coset of B in A). From an element a chosen from C and an element r in R we can form the coset $ra + B$ in A/B. This coset appears to depend on the choice of the element a from C. However, it actually depends only on C and r. For any (other) a' in C, $a' - a$ is in B; since B is a submodule of A, $r(a' - a) = ra' - ra$ is also in B. Since $ra' - ra$ is in B, $ra' + B = ra + B$. Thus, there is a well-defined operation of R on A/B, which may be denoted simply by $r(a + B) = ra + B$. I will demonstrate that this operation provides an R-module structure for A/B by verifying the four conditions listed in the definition of a module. Less experienced readers should read these verifications carefully, giving a precise reason at each equal sign.

(1) $r[(a + B) + (a' + B)] = r[(a + a') + B] = r(a + a') + B$
$= (ra + ra') + B = (ra + B) + (ra' + B) = r(a + B)$
$+ r(a' + B)$.

(2) $(r + r')(a + B) = (r + r')a + B = (ra + r'a) + B$
$= (ra + B) + (r'a + B) = r(a + B) + r'(a + B)$.

(3) $(rr')(a + B) = (rr')a + B = r(r'a) + B = r[(r'a) + B]$
$= r[r'(a + B)]$.

(4) $1(a + B) = 1a + B = a + B$.

The module A/B is called a quotient module.

If R is a field, then the quotient modules reduce to the familiar quotient spaces. If R is the ring Z of integers, then quotient modules may be regarded as indistinguishable from quotient groups. The role played by quotient modules within the general theory of modules will appear in § 9.

EXERCISE

1. Regard the ring R as a module over itself.
   a. Which subsets of R are submodules?
   b. For the special case in which R is a field F, describe the submodules and quotient modules of F.
   c. For the special case in which R is the ring Z of integers, describe the submodules and quotient modules of Z.

§ 3 DIRECT PRODUCTS AND DIRECT SUMS

Let R be a ring, and let $\{M_i \mid i \in I\}$ be a family of R-modules indexed by an arbitrary index set I. From this indexed family of modules we may form a single R-module, $\pi\{M_i \mid i \in I\}$, which we call *the direct product of* $\{M_i \mid i \in I\}$. The elements of $\pi\{M_i \mid i \in I\}$ are those functions $f : I \to \cup\{M_i \mid i \in I\}$ for which $f(j) \in M_j$ for each $j \in I$. The sum of elements $f, g \in \pi\{M_i \mid i \in I\}$ is defined by $(f + g)(i) = f(i) + g(i)$, and the scalar operation of an element $r \in R$ on an element $f \in \pi\{M_i \mid i \in I\}$ is defined by $(rf)(i) = rf(i)$. It is routine to verify that these definitions provide an R-module structure for $\pi\{M_i \mid i \in I\}$ (see Exercise 1a). Whenever no ambiguity seems likely, the module $\pi\{M_i \mid i \in I\}$ will be denoted in the contracted form πM_i. For I empty, we have (or define) $\pi M_i = 0$.

In our investigations the R-module $\pi\{M_i \mid i \in I\} = \pi M_i$ will be less important than one of its submodules. This submodule is denoted $\oplus\{M_i \mid i \in I\}$ and is called *the direct sum of* $\{M_i \mid i \in I\}$. The elements of $\oplus\{M_i \mid i \in I\}$ are those elements f of πM_i for which $f(j) = 0$ for all but finitely many $j \in I$. It is routine to verify that $\oplus\{M_i \mid i \in I\}$ is in fact a submodule of πM_i (Exercise 1b). When no ambiguity seems likely, $\oplus\{M_i \mid i \in I\}$ will be shortened to $\oplus M_i$. Notice that when I is finite $\oplus M_i = \pi M_i$.

My next objective is to clarify the sense in which products as defined above with an arbitrary index set may be regarded as a generalization of the elementary notion of the Cartesian product $M_1 \times M_2$, the elements of which are ordered pairs. Let $I = \{1, 2\}$. To fully describe an element $f \in \oplus\{M_i \mid i = 1, 2\} = \pi M_i$, it is sufficient to list the values that f assumes at 1 and 2, keeping in mind which value is associated with 1 and which with 2. An obvious way to do this is to list the *ordered* pair $(f(1), f(2))$. This procedure of associating $(f(1), f(2))$ with f provides a one–one function from $\oplus\{M_i \mid i = 1, 2\}$ into the Cartesian product set $M_1 \times M_2$. The procedure actually constitutes a one–one correspondence between $\oplus M_i$ and $M_1 \times M_2$, since for any ordered pair (m_1, m_2) in $M_1 \times M_2$ the function $g : I \to \cup M_i$ for which $g(1) = m_1$ and $g(2) = m_2$ is an element of $\oplus M_i$ and is associated with $(g(1), g(2)) = (m_1, m_2)$ by our process. Notice that for any $f, g \in \oplus M_i$ and $r \in R, f + g$ corresponds with $((f + g)(1), (f + g)(2)) = (f(1) + g(1), f(2) + g(2))$, and rf corresponds with $((rf)(1), (rf)(2)) = (rf(1), rf(2))$. The operations of $\oplus M_i$ therefore correspond to (familiar) coordinate-wise operations with ordered pairs. In view of these observations $\oplus\{M_i \mid i = 1, 2\}$ is normally written $M_1 \times M_2$ or $M_1 \oplus M_2$. The elements of $M_1 \times M_2 = M_1 \oplus M_2$ are then often thought of as ordered pairs that are added according to $(m_1, m_2) + (m'_1, m'_2) = (m_1$

$+ m_1', m_2' + m_2)$ and multiplied by elements $r \in R$ according to $r(m_1, m_2) = (rm_1, rm_2)$. Notice an additional convenience: if we have R-modules M and N, then $M \oplus N$ (or $M \times N$) has an obvious meaning. To grasp this same module with the function notation requires preparatory remarks: let $M_1 = M$, $M_2 = N$, $I = \{1, 2\}$, and consider $\pi\{M_i \mid i \in I\}$.

The discussion given for $I = \{1, 2\}$ can be carried out for $I = \{1, \ldots, n\}$ (replacing ordered pairs by ordered n-tuples); as a result, $\oplus\{M_i \mid i \in \{1, \ldots, n\}\} = \pi M_i$ is usually expressed in the form $M_1 \times \cdots \times M_n$ or $M_1 \oplus \cdots \oplus M_n$. For arbitrary (especially uncountably infinite) index sets I, it seems best to pass from the more elementary "tuple" approach to product sets to the function approach described above.

Direct sums will be fundamental to all our work. Basic properties of direct sums and direct products will be discussed in § 5 through § 8 after the prerequisite notion of homomorphism is introduced in the next section.

EXERCISE

1. a. Verify that for each family $\{M_i \mid i \in I\}$ of R-modules the set $\pi\{M_i \mid i \in I\}$ is an Abelian group under the addition defined for $f, g \in \pi M_i$ by $(f + g)(i) = f(i) + g(i)$ and becomes an R-module under the scalar operation defined for all $r \in R$ and $f \in \pi M_i$ by $(rf)(i) = rf(i)$.
   b. Verify that $\oplus M_i$ is a submodule of πM_i.

§ 4 HOMOMORPHISMS AND THEIR GRAPHS

Let A and B be modules over a ring R. A function $h : A \to B$ is called an *R-homomorphism* if $h(a + a') = h(a) + h(a')$ and $h(ra) = rh(a)$ hold for all a, a' in A and all r in R. When it is clear what ring is involved, we shorten R-homomorphism to homomorphism and more commonly to *map*.

When R is a field, the R-homomorphisms are just the familiar R-linear functions studied in linear algebra. If $h : A \to B$ is a homomorphism of an Abelian group A into an Abelian group B and if A and B are regarded as modules over the ring Z of integers as described in § 1, then h is automatically a Z-homomorphism. Thus, for Abelian groups (that is, Z-modules), the group homomorphisms and the Z-homomorphisms are the same functions. Without this fact we could not fully identify the theory of Abelian groups with the theory of Z-modules. Finally, here is a warning: be careful to notice the gross difference between module homomorphisms and ring homomorphisms (see Exercise 1). We will use module homomorphisms (maps) constantly, but in the body of the text we will use ring homomorphisms only in Observation 7 of Chapter 2 and its application in § 2 of Chapter 13.

Now let R be a ring and $h : A \to B$ be a map of the R-module A into the R-module B. The set $\{a \in A \mid h(a) = 0\}$ is called the *kernel* of h and is denoted $\ker(h)$. The kernel is a submodule of A (see Exercise 2a). If $\ker(h) = 0$, then h is said to be a *monomorphism.* A homomorphism has the property that distinct elements of its domain have distinct images (that is, a homomorphism is one–one) if and only if it is a monomorphism. The set $\{h(a) \mid a \in A\}$ is called the *image* of h and may be denoted $\mathrm{im}(h)$ or $h(A)$. The image is a submodule of B (see Exercise 2b). If $\mathrm{im}(h) = B$, then h is said to be an *epimorphism.* A homomorphism is surjective (that is, onto) if and only if it is an *epimorphism.* If h is both a monomorphism and an epimorphism (monic and epic), then it is said to be an *isomorphism.* In the special case in which $A = B$, we say that h is an *endomorphism* (of A); if, further, h is an isomorphism, we say that h is an *automorphism* (of A). We have seen that a module homomorphism is the source of at least two submodules—its kernel and image. A submodule S of an R-module M is also the source of an R-homomorphism: let $k : S \to M$ be the function defined for each s in S by $k(s) = s$, where on the right side of the equation we are recognizing that s is also an element of M. Then k is an R-monomorphism and is called *the inclusion map of S in M*. When $S = M$, k is also called *the identity map on M*. My next objective is to show that there is a way of regarding all module homomorphisms as special types of submodules—a fact that will prove to be useful in Chapters 6 and 7.

By the *graph* of a function $h : A \to B$, I mean the subset $\mathrm{gr}(h) = \{(a, h(a)) \mid a \in A\}$ of the Cartesian product set $A \times B$. When A and B are R-modules and h is a map, $A \times B$ may be regarded as an R-module (as described in § 3), and $\mathrm{gr}(h)$ then becomes a submodule of $A \times B$, since $(a, h(a)) + (a', h(a')) = (a + a', h(a) + h(a')) = (a + a', h(a + a'))$ and $r(a, h(a)) = (ra, rh(a)) = (ra, h(ra))$ are in $\mathrm{gr}(h)$ for all a, a' in A and all r in R. Distinct maps have distinct submodules as graphs. Which submodules of $A \times B$ are graphs of maps from A into B? To answer this question we need to observe one more property of $\mathrm{gr}(h)$: let $p : A \times B \to A$ be the function defined by $p(a, b) = a$ for each a in A and b in B. Since h is a (single-valued) function with domain A, the restriction, $p \mid \mathrm{gr}(h)$, of the function p to the subset $\mathrm{gr}(h)$ is a function from $\mathrm{gr}(h)$ into A that constitutes a one–one correspondence between $\mathrm{gr}(h)$ and A. This last property characterizes the submodules of $A \times B$ that are graphs of maps: *a sub-*

module S of $A \times B$ is the graph of an R-homomorphism $k : A \to B$ if and only if $p \mid S$ is a one–one correspondence between S and A. Given such a submodule S, it is easy to construct the required k: for each a in A, there is exactly one pair (a, b) in S, and $k(a) = b$. That k is a map follows from the fact that S is a submodule of $A \times B$ (see Exercise 3).

Sections 5, 6, and 8 deal with classes of module homomorphisms that are useful in the investigation of module structure. The general structure of homomorphisms is outlined in § 9. The function p used above is an example of a projection map as introduced in the next section.

EXERCISES

1. Regard the field C of complex numbers as a module over itself.
   a. Verify that the function $h : C \to C$ defined by $h(a + bi) = a - bi$ is a ring homomorphism but not a C-module homomorphism.
   b. Verify that the function $k : C \to C$ defined by $k(a + bi) = 2a + 2bi$ is a C-module homomorphism but not a ring homomorphism.
2. Let $h : A \to B$ be a homomorphism of R-modules.
   a. Verify that ker(h) is a submodule of A.
   b. Verify that im(h) is a submodule of B.
3. Let S be a submodule of the R-module $A \times B$ with the property that for each a in A there is a unique (a, b) in S. Verify that k is an R-homomorphism where the function $k : A \to B$ is defined for each a in A as that b in B for which (a, b) is in S.

§ 5 PROJECTIONS AND INJECTIONS (FOR DIRECT PRODUCTS)

Let $\{M_i \mid i \in I\}$ be a family of modules over a ring R, and let $\pi\{M_i \mid i \in I\}$ be the direct product of this family as defined in § 3. With each j in I we associate a function $p_j : \pi M_i \to M_j$ defined for each f in πM_i by evaluating f at j. Thus, $p_j(f) = f(j)$. The function p_j is an R-homomorphism since $p_j(f + f') = (f + f')(j) = f(j) + f'(j) = p_j(f) + p_j(f')$ and $p_j(rf) = (rf)(j) = rf(j) = rp_j(f)$ hold for all $f, f' \in \pi M_i$ and all $r \in R$. The map p_j is called the jth *projection map* associated with πM_i. My next objective will be to explain a commonly used construction involving products and their projection maps.

Let $\{h_i \mid i \in I\}$ be a family of R-homomorphisms $h_i : A \to M_i$ having the same domain A. From the family $\{h_i \mid i \in I\}$ we define a single function $h : A \to \pi M_i$ by associating with each a in A the function f_a in πM_i for which $f_a(i) = h_i(a)$ for each i in I. Thus, $h(a) = f_a$. To show that h is an R-homomorphism, we must show that, for all $a, a' \in A$ and $r \in R$,

$h(a + a')$ is the same function as $h(a) + h(a')$ and $h(ra)$ is the same function as $rh(a)$. To show that two functions are identical, we apply each to an arbitrary element of their common domain and verify that they have the same effect. Thus, $h(a + a') = h(a) + h(a')$ because for each i in I we have $h(a + a')(i) = f_{a+a'}(i) = h_i(a + a') = h_i(a) + h_i(a')$ and $(h(a) + h(a'))(i) = h(a)(i) + h(a')(i) = f_a(i) + f_{a'}(i) = h_i(a) + h_i(a')$. Likewise, $h(ra) = rh(a)$ since for each i in I we have $h(ra)(i) = f_{ra}(i) = h_i(ra) = rh_i(a)$ and $rh(a)(i) = rf_a(i) = rh_i(a)$. The modules and maps in this discussion can be displayed as a family of similar diagrams—one for each j in I (see Figure 1). *Each*

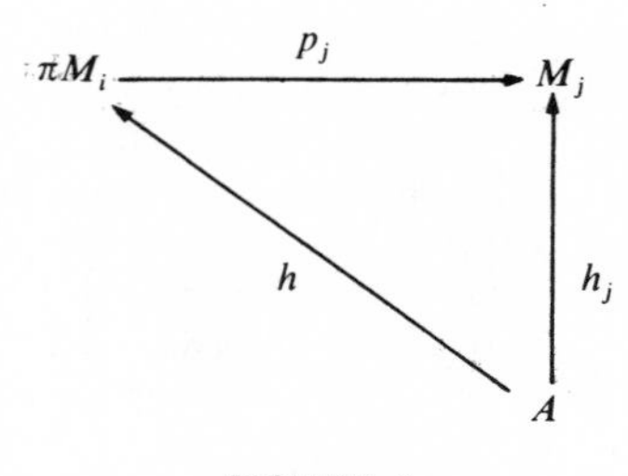

FIGURE 1

diagram is commutative in the sense that the two maps from A into M_j that are available in the diagram are equal, as we can easily verify: $p_j h(a) = p_j(f_a) = f_a(j) = h_j(a)$. In general, a *commutative diagram* is a diagram made up of modules and maps having the property that, for any pair of modules in the diagram, any two paths of maps in the diagram that connect the first module to the second must provide the same composite map. The map h constructed above is called the *product* of the family $\{h_i \mid i \in I\}$. The use of such a product map is illustrated in the proof of Observation 2 of Chapter 8.

There is a second family of maps associated with πM_i. For each j in I, let $q_j : M_j \to \pi M_i$ be the function for which, for each x in M_j, $q_j(x)$ is the function in πM_i that takes the value x at j and the value 0 elsewhere. Thus, $q_j(x)(j) = x$ and $q_j(x)(i) = 0$ for $i \neq j$. To show that q_j is a map, we must show that for all $x, x' \in M_j$ and $r \in R$, $q_j(x + x')$ is the same function as $q_j(x) + q_j(x')$ and $q_j(rx)$ is the same function as $rq_j(x)$. Since all four of these functions are zero except at j, the following computations are sufficient to verify that q_j is a map: $q_j(x + x')(j) = x + x' = q_j(x)(j) + q_j(x')(j) = (q_j(x) + q_j(x'))(j)$ and $q_j(rx)(j) = rx = r(q_j(x)(j)) = (rq_j(x))(j)$. The map q_j is called the jth *injection map* associated with πM_i.

Following the precise definition of πM_i, the M_j are not literally subsets of πM_i. Nevertheless, πM_i is not likely to be a valuable construction unless the modules M_j can in some sense be regarded as submodules of πM_i. The maps q_j are the formal vehicles that allow us to think of the M_j as being contained in πM_i. Precisely, $M_j \not\subset \pi M_i$; but $M_j \cong \mathrm{im}(q_j) \subset \pi M_i$. I have found that the simplicity of many discussions and proofs is obscured if I insist on maintaining the distinction between M_j and $\mathrm{im}(q_j)$. I therefore commonly regard each M_j as a subset of πM_i. Moreover, for each subset J of I, I regard $\pi\{M_i \mid i \in J\}$ as a subset of $\pi\{M_i \mid i \in I\}$. Any confusion arising over one of these pseudoinclusions can be resolved by factoring the pseudoinclusion into an isomorphism followed by an authentic inclusion.

With each product $\pi\{M_i \mid i \in I\}$ I have associated the two fundamental families of maps—the projections $\{p_i \mid i \in I\}$ and the injections $\{q_i \mid i \in I\}$. Each injection q_j can be composed with each projection p_k. Composites of the form $p_j q_j$ yield identity maps since, for each x in M_j, $(p_j q_j)(x) = p_j(q_j(x)) = q_j(x)(j) = x$. Composites of the form $p_k q_j (k \neq j)$ yield zero maps since, for each x in M_j, $(p_k q_j)(x) = p_k(q_j(x)) = q_j(x)(k) = 0$. The fact that $p_j q_j$ is the identity map of M_j encompasses two obvious but important facts: (1) the projections p_j are epimorphisms, and (2) the injections q_j are monomorphisms. In accord with the informal usage as described above, $\mathrm{im}(q_j) = M_j$ and $\ker(p_j) = \pi\{M_i \mid i \in I \setminus \{j\}\}$.

When $I = \{1, 2\}$, we may regard the elements of $\pi M_i = M_1 \oplus M_2$ as ordered pairs as explained in § 3. Our constructions and computations are especially transparent in this case. The projections and injections are exhaustively described by $p_1(m_1, m_2) = m_1$, $p_2(m_1, m_2) = m_2$, $q_1(m_1) = (m_1, 0)$, and $q_2(m_2) = (0, m_2)$, where m_1 and m_2 are aribitrary elements of M_1 and M_2, respectively. Each pair of maps $h_1 : A \to M_1$ and $h_2 : A \to M_2$ fits into a commutative diagram (Figure 2) where the product map h

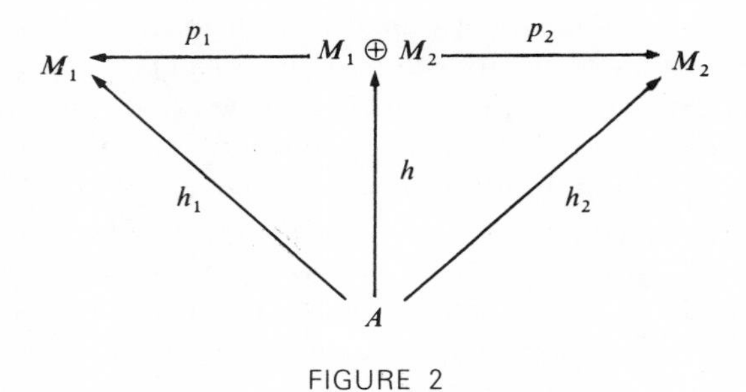

FIGURE 2

admits the simple description: $h(a) = (h_1(a), h_2(a))$ for each a in A. The informality of identifying M_i with $\mathrm{im}(q_i)$ amounts to blurring the distinction between M_1 and $\{(m_1, 0) \mid m_1 \in M_1\}$ and between M_2 and $\{(0, m_2) \mid m_2 \in M\}$. The computation of such composites as $p_1 q_1$ and $p_2 q_1$

can be illustrated by $(p_1q_1)(m_1) = p_1(m_1, 0) = m_1$ and $(p_2q_1)(m_1) = p_2(m_1, 0) = 0$.

§ 6 INJECTIONS AND PROJECTIONS (FOR DIRECT SUMS)

Let $\{M_i \mid i \in I\}$ be a family of modules over a ring R, let $\pi\{M_i \mid i \in I\}$ be the direct product of this family as defined in § 3, and let $\{p_i \mid i \in I\}$ and $\{q_i \mid i \in I\}$ be associated families of projections and injections as described in § 5. Inside πM_i we have the submodule $\oplus\{M_i \mid i \in I\}$. The restriction of each projection p_j to $\oplus M_i$ constitutes a map $\oplus M_i \to M_j$. Rather than denote this map by the cumbersome notation $p_j \mid \oplus M_i$, I simply denote it by p_j. Thus, I will write both $p_j : \pi M_i \to M_j$ and $p_j : \oplus M_i \to M_j$. No confusion will arise because the domains of the p_j will be clear from the contexts of our discussions. Since $\mathrm{im}(q_j) \subseteq \oplus M_i, q_j$ provides a map $M_j \to \oplus M_i$, and I denote this map by q_j also. Thus, I will write both $q_j : M_j \to \pi M_i$ and $q_j : M_j \to \oplus M_i$. The maps q_j taken in either sense are called injections, and the p_j taken in either sense are called projections. The system consisting of $\oplus M_i$ and its injections has a property that is closely related (diagrammatically dual) to the property we observed of πM_i and its projections in § 5.

Let $\{k_i \mid i \in I\}$ be a family of R-homomorphisms $k_i : M_i \to A$, where A is a fixed R-module. From the family $\{k_i \mid i \in I\}$ we define a single function $k : \oplus M_i \to A$ by associating with each f in $\oplus M_i$ the sum of the following (finite) set of elements of $A : \{k_j(f(j)) \mid f(j) \neq 0\}$. The following convention will be used throughout this book: *if all but finitely many members of an indexed family of elements of an Abelian group A are zero, then by the sum of the family we will mean the sum of the nonzero members of the family.* This convention allows us to describe k by the equation $k(f) = \Sigma\{k_i f(i) \mid i \in I\}$ or, in abbreviated form, $\Sigma k_i f(i)$. Two computations show that k is an R-homomorphism: $k(f + f') = \Sigma k_i(f + f')(i) = \Sigma k_i(f(i) + f'(i)) = \Sigma(k_i f(i) + k_i f'(i)) = \Sigma k_i f(i) + \Sigma k_i f'(i) = k(f) + k(f')$ and $k(rf) = \Sigma k_i(rf)(i) = \Sigma k_i(rf(i)) = \Sigma r k_i f(i) = r\Sigma k_i f(i) = rk(f)$ hold for all $f, f' \in \oplus M_i$ and all $r \in R$. The modules and maps in our discussion provide a family of similar diagrams, one for each j in I (see Figure 3). *Each diagram is commutative* since for each x in M_j we have $kq_j(x) = \Sigma\{k_i(q_j(x)(i)) \mid i \in I\}$ and since $q_j(x)$ assumes a nonzero value only at j, $kq_j(x) = k_j q_j(x)(j) = k_j(x)$. The usefulness of maps constructed in the manner of this map k will be illustrated in § 7.

For the remainder of this section we will consider the special case of the direct sum of a pair of modules and introduce at the same time a

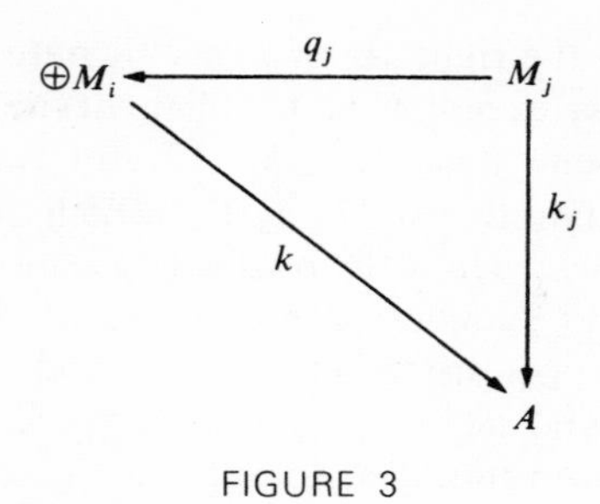

FIGURE 3

convenient variation of notation. Let A and B be R-modules. The projections of $A \oplus B$ onto A and B will be denoted by p_A and p_B, and the injections of A and B into $A \oplus B$ by q_A and q_B. Thus, we have, for example, $p_A(a, b) = a$ and $q_B(b) = (0, b)$ holding for all a in A and b in B. Each pair of maps $h_A : A \to C$ and $h_B : B \to C$ fits into a commutative diagram (Figure 4) where h admits the simple description $h(a, b) = h(a) + h(b)$.

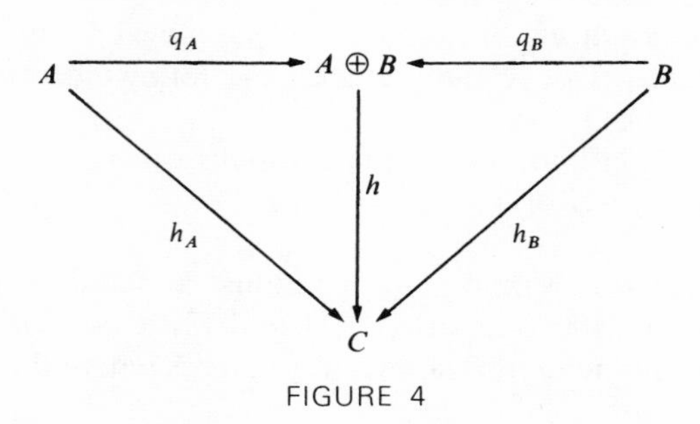

FIGURE 4

The closing remarks of § 4 concerning graphs of maps may be reformulated in our present terminology: a submodule S of $A \oplus B$ is the graph of a map $k : A \to B$ if and only if the restriction, $p_A \mid S : S \to A$, of p_A to S is an isomorphism. A submodule T of $A \oplus B$ for which $p_A \mid T$ is monic may be regarded as the graph of a map from $p_A(T)$ into B.

§ 7 INTERNAL DIRECT SUMS

A constant theme of our work will be the attempt to represent an R-module A of some specified type as a direct sum of modules of some structurally more transparent type. With out *present* meaning of direct sum, the appropriate question for us to ask would *not* be "Is there a family of R-modules $\{M_i \mid i \in I\}$ for which $A = \oplus M_i$?" because the elements of the set on the right are of a specific nature: they are functions from I into $\cup M_i$. Thus (except when the elements of A are such functions), A has no chance of being *equal* to $\oplus M_i$. Our question would have to be phrased "Is there a family $\{M_i \mid i \in I\}$ for which $A \cong \oplus M_i$?" My next objective will be to observe how the problem of establishing isomorphisms such as $A \cong \oplus M_i$ can be converted into an "internal" problem of looking within A itself for an appropriate family of submodules. Once the observation has been made and its consequences have been discussed, our terminology will have been modified to allow us to restore the question to its original form: "Is there a family of (sub)modules $\{M_i \mid i \in I\}$ for which $A = \oplus M_i$?"

OBSERVATION 1. An R-module A is isomorphic with a direct sum $\oplus\{M_i | i \in I\}$ if and only if A contains a family $\{A_i | i \in I\}$ of submodules that satisfies the following conditions:

(1) For each $i \in I$, $A_i \cong M_i$.
(2) For each $a \in A$ there is a *unique* indexed family $\{a_i \mid i \in I\}$ of elements of A that satisfies the following three conditions:
  (a) $a_i \in A_i$ for each i in I.
  (b) $a_i = 0$ for all but finitely many i in I.
  (c) $a = \Sigma\{a_i \mid i \in I\}$.

Verification: Suppose A contains a family of submodules $\{A_i \mid i \in I\}$ that satisfies conditions (1) and (2). Consider the family of diagrams (one for each j in I) shown in Figure 5, where the maps k_j and k

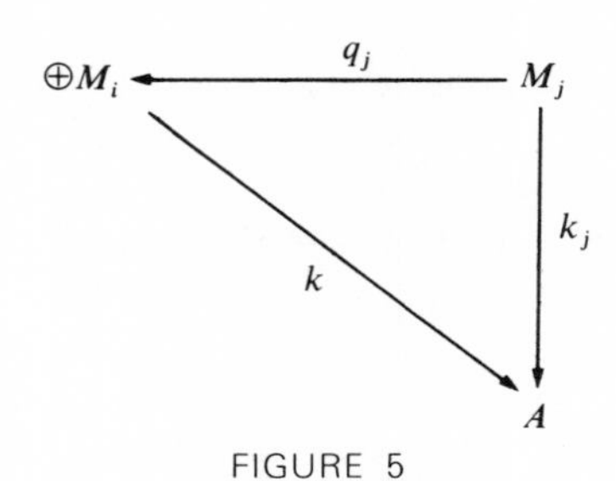

FIGURE 5

are constructed as follows. Map k_j is the composite of an isomorphism $M_j \to A_j$ (the existence of which is guaranteed by condition (1)) with the inclusion map $A_i \subseteq A$. Map k is defined by $k(f) = \Sigma k_i f(i)$ (see § 6).

We will show that k is an isomorphism. Let f be in ker(k). Then $0 = \Sigma k_i f(i)$. We will use the uniqueness assumed in condition (2) as follows. Let $\{z_i \mid i \in I\}$ be the identically zero indexed family; that is, $z_i = 0$ for each i in I. Each of the two families $\{k_i f(i)\}$ and $\{z_i\}$ is finitely nonzero and has sum 0. By condition (2), they must be identical as indexed families. Thus, $k_j f(j) = 0$ for each j in I. Since each k_j is a monomorphism, $f(j) = 0$ for each j in I. Thus, f is the zero element of $\oplus M_i$, and k is monic. To show that k is epic, we let a be an arbitrary element of A and construct an f in $\oplus M_i$ as follows. By condition (2) we have $a = \Sigma a_i$, and, since each a_i is in $\text{im}(k_i)$, we also have $a = \Sigma a_i = \Sigma k_i(x_i)$ for elements x_i in M_i. (Moreover, all but finitely many of the x_i are zero.) Let f be the function in $\oplus M_i$ for which $f(j) = x_j$ for each j in I. Then $k(f) = \Sigma k_i f(i) = \Sigma k_i(x_i) = a$ as required. We have shown that k is an isomorphism.

Suppose now that $A \cong \oplus M_i$. Our argument will be transparent if we treat first the special case in which $A = \oplus M_i$. In this case it is easily verified that conditions (1) and (2) are satisfied by the family defined by $A_i = q_i(M_i)$ (see Exercise 1a). Passing to the general case, let $h : \oplus M_i \to A$ be a specific isomorphism. Then the family defined by $A_i = (hq_i)(M_i)$ can be verified to satisfy conditions (1) and (2) (see Exercise 1b).

An R-module A is said to be the *internal direct sum* of a family $\{A_i \mid i \in I\}$ of submodules if the family satisfies condition (2) of Observation 1. When it is necessary to distinguish carefully between the new notion of an *internal* direct sum and the previously defined notion of a direct sum, the latter may be called the *external* direct sum. I will summarize the relation between the internal and external direct sum concepts.

Let A be the internal direct sum of submodules $\{A_i \mid i \in I\}$. Then the external direct sum $\oplus A_i$ can be formed and the observation asserts that A is isomorphic with this external direct sum. On the other hand, we may start with an arbitrary family of modules $\{A_i \mid i \in I\}$ and form the external direct sum $\oplus A_i$. This external direct sum is the internal direct sum of the family $\{q_i(A_i) \mid i \in I\}$ and $q_i(A_i) \cong A_i$ for each i in I.

In view of the close relationship between internal direct sums and external direct sums, I will use the same notation to denote both. *When an R-module A is the internal direct sum of submodules* $\{A_i \mid i \in I\}$, *I will write* $A = \oplus\{A_i \mid i \in I\}$. The ambiguity of the notation presents no problem. When the A_i are literally submodules of A, the internal direct sum is intended; otherwise, the external direct sum is intended. (When the informality of identifying each $\text{im}(q_i)$ with A_i in the external direct sum is resolutely maintained, it leads to the identification of the internal and external direct sums.)

I will often be interested in showing that a module is the (internal) direct sum of two of its submodules. A slight variation in procedure is possible in this case (see Exercise 2): for submodules B and C of an

R-module A, $A = B \oplus C$ *if and only if* $B \cap C = 0$ *and for each* a *in* A *there are elements* b *in* B *and* c *in* C *for which* $a = b + c$.

EXERCISES

1. a. Show that for each family $\{M_i \mid i \in I\}$ of R-modules, the external direct sum $\oplus\{M_i \mid i \in I\}$ is the internal direct sum of the family of submodules $\{\text{im}(q_i) \mid i \in I\}$.
   b. Let $h : A \to B$ be an isomorphism of R-modules and suppose that A is the internal direct sum of submodules $\{A_i \mid i \in I\}$. Show that B is the internal direct sum of the submodules $\{h(A_i) \mid i \in I\}$.
2. Let B and C be submodules of an R-module A. Show that A is the internal direct sum of B and C if and only if $B \cap C = 0$ and for each a in A there are elements b in B and c in C for which $a = b + c$.

§ 8 SPLITTING MAPS AND SUMMANDS

The projection p_A and the injection q_A associated with a direct product $A \oplus B$ have the following two properties: the composite map $p_A q_A$ is the identity map on A, and $A \oplus B$ is the internal direct sum of $\text{im}(q_A)$ and $\ker(p_A)(= \text{im}(q_B))$. A comparison of these two properties of p_A and q_A motivates the following observation.

OBSERVATION 2. Let $h : U \to V$ and $k : V \to U$ be homomorphisms between R-modules U and V for which kh is the identity map on U. Then $V = \text{im}(h) \oplus \ker(k)$.

Verification: Let $x \in \text{im}(h) \cap \ker(k)$. Then $x = h(u)$ for some $u \in U$ and also $k(x) = 0$. Consequently, $u = kh(u) = k(x) = 0$ and $x = h(u) = h(0) = 0$. Thus, $\text{im}(h) \cap \ker(k) = 0$. Let $v \in V$. Then $hk(v) \in \text{im}(h)$ and $v = hk(v) + (v - hk(v))$. To complete the verification, we need only show that $v - hk(v)$ is in $\ker(k)$: $k(v - hk(v)) = k(v) - khk(v) = k(v) - k(v) = 0$.

When maps $h : U \to V$ and $k : V \to U$ have the property that kh is the identity map, we say that each of the maps is a *splitting map* for the other. As our observation shows, splitting maps are valuable tools for splitting modules into internal direct sums. Notice that, when kh is the identity map, it follows that h is monic and k is epic. The maps p_A and q_A mentioned above are splitting maps for each other.

A submodule B of an R-module A is a *summand* of A if there is a submodule C of A for which $A = B \oplus C$. When such a C exists, it is said to be a *complementary summand* for B. For an arbitrary module A, $A = A \oplus 0$, and consequently A and the zero submodule are complementary summands for each other. Observation 2 provides the following useful strategies for verifying that certain submodules are summands. If $h : A \to B$ is a monomorphism and we wish to show that im(h) is a summand of B, then it is sufficient to find a map $s : B \to A$ for which $sh(a) = a$ for all a in A. If $h : A \to B$ is an epimorphism and we wish to show that ker(h) is a summand of A, then it is sufficient to find a map $s : B \to A$ for which $hs(b) = b$ for all b in B. Finally, if A is a submodule of B and we wish to show that it is a summand, then it is sufficient to find a map $s : B \to A$ for which $s(a) = a$ for all a in A.

A summand S of an R-module M may have many distinct complementary summands (see Exercises 1 and 2). There is a one–one correspondence between the set of complementary summands of S and the set of splitting maps for the inclusion map of S in M (see Exercise 3). The complementary summands for the summand im(q_B) of the external direct sum $A \oplus B$ are precisely the graphs of the homomorphisms of A into B (see Exercise 4).

EXERCISES

1. Let A be a noncyclic Abelian group (that is, a Z-module) of order 4. Show that A is the direct sum of any two of its three cyclic submodules of order 2.
2. Let R be the ring Z_2 of integers modulo 2. Each submodule of order 2^n of each R-module of order 2^{n+1} is a summand having precisely 2^n distinct complementary summands.
3. Let S be a summand of an R-module M, and let $h : S \to M$ be the inclusion map of S into M. Show that the function associating each splitting map $k : M \to S$ of h with ker(k) is a one–one correspondence between the set of splitting maps of h and the set of all complementary summands of S in M.
4. Show that the complementary summands of the submodule im(q_B) of the R-module $A \oplus B$ are precisely the graphs of the homomorphisms of A into B.

§ 9 NATURAL MAPS

Let B be a submodule of an R-module A, and let A/B be the associated quotient module as described in § 2. Since each element of A lies in precisely one coset of B, we may define a function $n : A \to A/B$ by specifying that, for each a in A, $n(a)$ is that coset to which a belongs. In the usual coset notation this function is described by $n(a) = a + B$. Since $n(a + a') = a + a' + B = a + B + a' + B = n(a) + n(a')$ and $n(ra) = ra + B = r(a + B) = rn(a)$ hold for all a, a' in A and all r in R, n is an R-homomorphism. The map n is called the *natural map* of A into A/B.

Using natural maps (and inclusion maps), we can analyze arbitrary maps in a revealing way. For an arbitrary map $h : A \to B$ between R-modules A and B, Figure 6 is constructed. All modules and

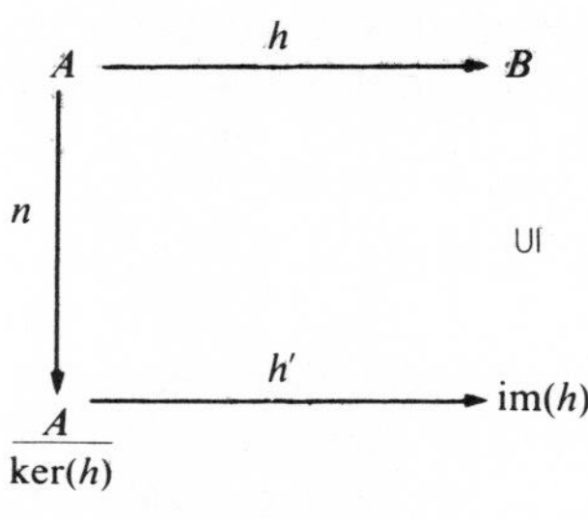

FIGURE 6

maps in the diagram have an established meaning except h'. To define h', we proceed as follows: if $a + \ker(h) = a' + \ker(h)$, then $a - a' \in \ker(h)$ and consequently $0 = h(a - a') = h(a) - h(a')$. Thus, $h(a) = h(a')$. As a result, the following equation defines a function from $A/\ker(h)$ into $\mathrm{im}(h)$: $h'(a + \ker(h)) = h(a)$. It is easy to verify that h' is an R-isomorphism (see Exercise 1). Since $h'n(a) = h'(a + \ker(h)) = h(a)$, the diagram is commutative.

This commutative diagram embodies a great deal of interesting information about the nature of homomorphisms. Here are three easily quoted immediate consequences: (1) If $h : A \to B$ is an epimorphism, then $B \cong A/\ker(h)$. (2) Every map can be factored into the composite of three maps with very special features—that is, a natural map followed by an isomorphism followed by an inclusion map. (3) Every map can be expressed as an epimorphism followed by a monomorphism. Perhaps the central significance of the diagram is that it reduces the problem of determining *all* maps of an R-module A into an R-module B to the following program: determine all the submodules of A and B; for each pair of submodules K of A and S of B, determine all isomorphisms h' that exist (if any) between A/K and S. Each triple (K, h', S) for which $h' : A/K \to S$ is an isomorphism then yields a map $h : A \to B$, and every map from A into B arises in this way precisely once. As for the cardinal number of the set of maps from A into B, this can be expressed in the especially simple form $\Sigma\{v(S)\alpha(S) \mid S$

is submodule of $B\}$, where $\alpha(S)$ is the cardinal number of the set of automorphisms of the module S and $\nu(S)$ is the cardinal number (possibly 0) of the set consisting of those submodules K of A for which A/K is isomorphic with S.

EXERCISES

1. Show that the function $h' : A/\ker(h) \to \mathrm{im}(h)$ constructed in §9 is an R-isomorphism.
2. Show that, if B is a summand of an R-module A and if C_1 and C_2 are complementary summands of B in A, then $C_1 \cong A/B \cong C_2$.
3. Let B and C be submodules of an R-module A.
   a. Verify that $B \cap C$ is a submodule of A.
   b. Verify that the subset $B + C$ defined by $B + C = \{b + c \mid b \in B, c \in C\}$ is a submodule of A.
4. Let B and C be submodules of an R-module A, and let $n : A \to A/C$ be the natural map. Let $h : B + C \to A/C$ be the restriction of n to $B + C$, and let $k : B \to A/C$ be the restriction of n to B.
   a. Find the kernels and images of h and k.
   b. Show that $(B + C)/C \cong B/(B \cap C)$.
5. Let $\{A_i \mid i \in I\}$ be a family of R-modules, and, for each $i \in I$, let B_i be a submodule of A_i. Show that $(\oplus A_i)/(\oplus B_i) \cong \oplus(A_i/B_i)$.

§ 10 SUBSETS AND THE SUBMODULES THEY GENERATE

Let A be a module over a ring R. Given a single element s in A, what can we build? We can multiply s by each scalar r in R to produce the set $\{rs \mid r \in R\}$, which will be denoted Rs. It is mentally verifiable that Rs is a submodule of A and, since $s = 1s$, contains s. Since any submodule of A that contains s must contain Rs, we observe that there exists a unique smallest submodule of A containing s, and Rs is this smallest submodule. We will say that Rs is the submodule of A generated by s (or by $\{s\}$). Given two elements s_1, s_2 of A, what can we build? We can build Rs_1 and Rs_2, and from these submodules we can form the set of all sums $\{r_1s_1 + r_2s_2 \mid r_1, r_2 \in R\}$, which will be denoted $Rs_1 + Rs_2$. It is mentally verifiable that $Rs_1 + Rs_2$ is a submodule of A that contains both s_1 and s_2. Since any submodule of A that contains both s_1 and s_2 must contain $Rs_1 + Rs_2$, we observe that there exists a unique smallest submodule of A containing $\{s_1, s_2\}$, and $Rs_1 + Rs_2$ is this smallest submodule. We will say that $Rs_1 + Rs_2$ is the submodule of A generated by $\{s_1, s_2\}$.

The discussions we have given of the submodules generated by $\{s\}$ and by $\{s_1, s_2\}$ apply to any finite subset $\{s_1, \ldots, s_n\}$ of A. The submodule of A generated by $\{s_1, \ldots, s_n\}$ is $Rs_1 + \cdots + Rs_n = \{r_1 s_1 + \cdots + r_n s_n \mid r_1, \ldots, r_n \in R\}$, which is the unique smallest submodule containing $\{s_1, \ldots, s_n\}$. For a countably infinite subset of A, little notational change is needed. If we denote the set by $\{s_i \mid i \text{ a positive integer}\}$, then the submodule it generates is $Rs_1 + \cdots + Rs_i + \cdots = \{r_1 s_1 + \cdots + r_n s_n \mid n \text{ a positive integer and } r_1, \ldots, r_n \in R\}$. To give a convenient description of the submodule generated by an arbitrary subset S of A, it is perhaps best to formally pass over to finitely nonzero indexed families of scalars: *the submodule of A generated by S is* $\{\Sigma r_s s \mid s \in S, r_s = 0$ for all but finitely many $s \in S\}$. The submodule generated by the empty set is the zero submodule. The submodule of A generated by S is the unique smallest submodule of A that contains S; it is contained in every submodule of A that contains S.

Let A be an R-module. Then A is said to be *cyclic* if there is an element s of A for which $A = Rs$. Likewise, A is *finitely generated* if there is a finite subset $\{s_1, \ldots, s_n\}$ of A for which $A = Rs_1 + \cdots + Rs_n$ and A is *countably generated* if either A is finitely generated or $A = Rs_1 + \cdots + Rs_i + \cdots$ for some countable subset $\{s_i \mid i \text{ a positive integer}\}$ of A.

For a first set of examples, let A be a three-dimensional vector space over the field R of real numbers. If $s = 0$, the submodule (that is, subspace) generated by s is the zero submodule; if $s \neq 0$, the submodule is the line containing s and the origin. The submodule generated by $\{s_1, s_2\}$ is $Rs_1 + Rs_2$, which is (1) 0 if $s_1 = 0 = s_2$, (2) the plane containing s_1, s_2 and the origin if s_1, s_2 and the origin are not contained in a common line, and (3) the line containing s_1, s_2 and the origin in all other cases. Every submodule of A is finitely generated (by three or fewer elements).

For a second set of examples, we will let A be the additive group (that is Z-module) of real numbers. The submodule (that is, subgroup) generated by $\{1/2, 2/3\}$ is $Z(1/2) + Z(2/3)$, which is actually a cyclic submodule since it is identical with $Z(1/6)$. The submodule generated by $\{1, \sqrt{2}\}$ is $Z + Z\sqrt{2} = Z \oplus Z\sqrt{2}$. The submodule Q consisting of the rational numbers is not finitely generated since for each finite subset $\{n_1/d_1, \ldots, n_k/d_k\}$, where the n_i and d_i are integers, we have $Z(n_1/d_1) + \cdots + Z(n_k/d_k) \subseteq Z(1/(d_1 \ldots d_k))$. One of the more interesting generating sets for Q is $\{1/n! \mid n \text{ a positive integer}\}$; every *infinite* subset of this generating set is also a generating set for Q.

The approach we have given to submodules generated by subsets might be thought of as an approach from the inside. An equivalent approach from the outside can be based on the important (and mentally verifiable) fact that the intersection of any (nonempty) family of submodules of an R-module A is again a submodule of A. For any subset S of an R-module A, the intersection $\cap\{A_i \mid i \in I\}$ of all those submodules A_i of A that contain S is a submodule of A containing S. Since each submodule of A that contains S is one of the A_i, it is clear that $\cap A_i$ is the unique smallest submodule that contains S. Thus, $\cap A_i$ is the submodule of A generated by S. This outside approach is attractive but for working purposes the inside approach is usually more useful.

Cyclic modules will play an important role throughout our work. Finitely generated modules receive close attention in Chapters 10, 11, and 13. Countably generated modules are constantly under study in Part Three. Some attention to generating sets in general is given in Chapter 5.

§ 11 MAXIMAL NESTS OF SUBSETS

Let S be a set and let $\mathscr{F}$ be a family of subsets of S. A subfamily $\mathscr{N}$ of $\mathscr{F}$ is a *nest* in $\mathscr{F}$ if for each pair of sets $A \in \mathscr{N}$ and $B \in \mathscr{N}$ we have either $A \subseteq B$ or $B \subseteq A$. A nest $\mathscr{M}$ in $\mathscr{F}$ is a *maximal nest* in $\mathscr{F}$ if the only nest $\mathscr{M}'$ in $\mathscr{F}$ for which $\mathscr{M}' \supseteq \mathscr{M}$ is $\mathscr{M}' = \mathscr{M}$ itself. In some families of subsets it is possible to describe one or more specific maximal nests (see exercises). The situations in which we will use maximal nests are not usually of this sort. In fact, we will be assured of the existence of the maximal nests we need only by employing the following broad principle, which we will regard as an axiom of set theory: *for each set S and each family $\mathscr{F}$ of subsets of S, there exists at least one maximal nest in $\mathscr{F}$*. If you would like to see now how this principle is used, read the proof of Observation 2 in Chapter 1 or the Lemma in Chapter 2.

We hope that the maximal-nest principle will seem intuitively plausible and we believe that it should be used unself-consciously. It is one of several equivalent and widely used set-theoretic principles (the choice axiom, the well-ordering principle, Zorn's lemma, etc.). To study these set-theoretic principles now would only inhibit your progress in understanding the structure of modules. Later you may wish to examine Halmos [1960], Chapter 0 of Kelley [1955], or even Rubin and Rubin [1963].

EXERCISES

1. Let Q be the set of rational numbers and let $\mathscr{F}$ be the family of all finite subsets of Q. Describe all of the maximal nests $\mathscr{M}$ in $\mathscr{F}$. State precisely which subsets of Q are of the form $\cup\mathscr{M}$ for some maximal nest $\mathscr{M}$ in $\mathscr{F}$.

2. Let Q be the set of rational numbers. Let $\mathscr{F}$ be the family of all nonempty proper subsets of Q. For each x in Q, let $U_x = \{y \in Q \mid x < y\}$ and $C_x = \{y \in Q \mid x \leqq y\}$. Let $\mathscr{M}$ be the family consisting of the sets U_x for all x in Q, and the sets C_x for all x in Q. Is $\mathscr{M}$ a maximal nest in $\mathscr{F}$?

3. In Exercise 2, replace the set Q of rational numbers by the set R of real numbers. Is $\mathscr{M}$ a maximal nest in this case?

APPENDIX 1: MODULE THEORY WITHIN MATHEMATICS

In the notes to Chapters 4 and 5 I have indicated how major aspects of *linear algebra* may be seen as special aspects of module theory. For further clarification of the relationship of module theory to linear algebra, I suggest Hartley and Hawkes [1970]. Linear algebra provides one of the bridges across which module theory relates to the nonmathematical world. The applications of linear algebra to the physical sciences and human studies are too numerous to list, but I would like to mention the work by Kalman (see Kalman, Falb, and Arbib [1969]), which expresses a concern for a module-theoretic viewpoint in systems theory.

The theory of *Abelian groups* is equivalent to the theory of modules over the ring Z of integers and is therefore subsumed by module theory. A fundamental tool for the study of arbitrary groups is *group-representation theory*. In the notes to Chapter 3 I have indicated how representations of groups can be studied by means of modules. For a broader view of group-representation theory seen from within module theory, I suggest Burrow [1965] or Curtis and Reiner [1962]. Group-representation theory is widely studied and applied by physicists (see Boerner [1963]) and chemists (see Cotton [1971] or Hall [1969]) and consequently represents a second bridge relating module theory to nonmathematical concerns.

Among the structures discussed in introductory courses in abstract algebra, those most intimately related to modules are *rings*. The interplay between modules and rings is a major source of the dynamics of this book. The relation between modules and rings can be indicated by a ratio-proportion mnemonic: modules are to rings as permutation groups are to groups. If this remark seems mysterious, see page 15 of Hilton [1971] for an indication of the way in which a module may be regarded as a ring representation.

In *homological algebra* elaborate diagrams of modules and maps are constructed as tools for drawing conclusions about various algebraic objects including groups, rings, and fields. One of the purposes of this book is to provide a tool for testing the following hypothesis: homological algebra can be studied with greater pleasure and less frustration after a deliberate study of module structure.

In *algebraic topology* numerous groups are associated with each topological space; these groups are in some cases Abelian groups having natural module structures over rings other than the integers. Thus, the study of topological spaces provides examples of modules and motivation for their study.

In *categorical algebra* one of the major topics is Abelian categories. One of the outstanding theorems of categorical algebra, the Mitchell–Freyd theorem (see Freyd [1964]), can be paraphrased as follows: the relationships holding in a small Abelian category are the same as those holding among the modules over an appropriate ring. At this level of generality, systems of modules (and maps) have provided the concrete environments used to represent structures of a higher level of abstraction.

APPENDIX 2: RIGHT MODULES

Our definition of a left R-module A in § 1 was based on a function $m : R \times A \to A$. If we reverse the order of R and A in the product $R \times A$ and make those further notational changes that seem natural, we obtain the definition of a right R-module.

DEFINITION. A *right R-module* consists of an Abelian group A together with a function $m : A \times R \to A$, which satisfies the following conditions:

(1) $m(a + a', r) = m(a, r) + m(a', r)$,
(2) $m(a, r + r') = m(a, r) + m(a, r')$,
(3) $m(a, rr') = m(m(a, r), r')$, and
(4) $m(a, 1) = a$,

where r, r' are arbitrary elements of R and a, a' are arbitrary elements of A.

As in the case of left R-modules, the element $m(a, r)$ is conventionally denoted $a \cdot r$ or ar. Using the juxtaposition notation, the four conditions in the definition may be written as follows:

(1) $(a + a')r = ar + a'r$,
(2) $a(r + r') = ar + ar'$,
(3) $a(rr') = (ar)r'$, and
(4) $a1 = a$.

It might appear that right and left modules differ only notationally and not mathematically. This would be true except for the presence of condition (3), which constitutes a genuine mathematical distinction: for right modules the result of applying the product of two scalars to a module element is the same as the effect of applying the *first* scalar and then the *second*; whereas, for left modules, the effect of applying the product of two scalars is the same as applying the *second* scalar and then the *first*. Nevertheless, it is still possible to convert right modules into left modules (and left into right) but only at the cost of changing the ring of scalars.

With each ring R we associate a ring R°, called *the ring opposite to* R. As Abelian groups, R° and R are identical. The product of two elements r, r' of R° will be denoted $r * r'$ and be defined by $r * r' = r' \cdot r$, where $r' \cdot r$ is the product of r' and r in R. Let A be a right R-module. We will convert A into a left R°-module by defining $r \cdot a$ to be $a \cdot r$. The third condition in the definition of a left R°-module is verified as follows: $(r * r')a = a(r * r') = a(r' \cdot r) = (ar')r = (r'a)r = r(r'a)$. The remaining three conditions are satisfied on even more elementary grounds. A commutative ring is identical with its opposite and for such a ring no distinction between left and right modules is necessary.

We will study left modules only. A separate study of right modules is not necessary since each of our results on left modules has a natural reinterpretation as a result on right modules. I offer two paths of justification for this assertion: (1) Each result on left R-modules may be interpreted as a result on right R°-modules. (2) If appropriate left-right alterations were made throughout the present book, it would become a study of right modules with all results and proofs retaining their present form (aside from the left-right alterations).

CHAPTER 1

THE RING AS A MODULE OVER ITSELF

§ 1 INTRODUCTION

The backbone of this book is a sequence of ten theorems that describe the structure of certain classes of modules. Each of these theorems is a description of how the modules in question may be constructed by means of at most four processes: (1) choosing a submodule of R, (2) forming a quotient module of R, (3) building certain extensions of these quotients (to be explained in Chapter 7), and (4) forming direct sums. *Thus, R regarded as an R-module is the foundation for this book.* Since this module is so important, we will go carefully through its definition again. For the Abelian group we use R with its ring addition. For the scalar multiplication of the ring R on the Abelian group R, we use the ring multiplication. Reviewing the module axioms, we have the following: (1) $r(r' + r'') = rr' + rr''$ by the left distributivity of ring multiplication, (2) $(r + r')r'' = rr'' + r'r''$ by the right distributivity of ring multiplication, (3) $(rr')r'' = r(r'r'')$ by the associativity of ring multiplication, and (4) $1r = r$ by the definition of an identity element of a ring.

In § 2 we will discuss the submodules and quotient modules of R, which are basic building blocks in module theory. The identity element of R is a fundamental tool in the theory of R-modules. In § 3 I will show how this identity element can be used to produce strong results about mapping properties relating to R. In the process, you will make your first acquaintance with the concept of a projective module. This concept of projectivity is one of the most fundamental general concepts in module theory—perhaps the most fundamental.

§ 2 MODULES ASSOCIATED WITH R

We will examine the submodules, quotient modules, and direct summands of the ring R regarded as a module over itself.

Submodules of R. A subset of a ring R is a submodule of R if it is an additive subgroup of R and is closed under the scalar operation, which is multiplication on the left by elements of R. Thus, the conditions under which a subset of a ring is a submodule are precisely those conditions under which it is a left ideal. We conclude that *the submodules of R are precisely the left ideals of R.*

Quotient Modules of R. As an R-module, R is generated by the identity element 1. Each quotient R/L, where L is an arbitrary submodule (that is, left ideal) of R, is generated by the coset $1 + L$. This property of being cyclic is a characterization of the quotient modules of R: a module is isomorphic to a quotient module of R if and only if it is cyclic. This fact may be seen in the larger context of Observation 1, which will prove to be a basic tool for the study of modules.

OBSERVATION 1. For each element m of each R-module M, there is one and only one map $h : R \to M$ for which $h(1) = m$. The image of h is the cyclic submodule of M generated by m.

Verification: For m in M, define a function $h : R \to M$ by $h(r) = rm$. We have $h(r + r') = (r + r')m = rm + r'm = h(r) + h(r')$, $h(rr') = (rr')m = r(r'm) = rh(r')$, and $h(1) = 1m = m$. Thus, h satisfies the requirements of the observation. Let $k : R \to M$ be any map for which $k(1) = m$. Then $k(r) = k(r1) = rk(1) = rm = h(r)$. Thus, h is the only map meeting the requirements of Observation 1. The image of h is $\{h(r) | r \in R\} = \{rm | r \in R\} = Rm$, which is the cyclic submodule generated by m.

Two special types of cyclic modules will play basic roles in our studies—simple modules and uniform cyclic modules. I will introduce these modules and discuss them briefly.

DEFINITION. An R-module is *simple* if it is not zero and has only two submodules (itself and zero).

A simple module is automatically cyclic. In fact, it is cyclic in a rather colossal way: *each nonzero element s of a simple R-module S generates S.* Since s is not zero, the cyclic submodule it generates is not zero and therefore, by the simplicity of S, must be all of S.

The presence of the identity $(1 \neq 0)$ in a ring forces the existence of at least one simple module. In proving this fact I will also show that every such ring contains a maximal submodule (that is, left ideal) as stated in the following definition.

DEFINITION. A submodule M of a module A is a *maximal* submodule if it is proper and the only submodules B satisfying $M \subseteq B \subseteq A$ are M and A.

OBSERVATION 2. Each ring R (with $1 \neq 0$) contains at least one maximal left ideal M, and, for any such M, R/M is a simple module.

Verification: Consider the collection of *proper* left ideals of R. This collection of subsets of R must contain at least one maximal nest. Let $\{L_i \mid i \in I\}$ be such a nest. Let $M = \cup L_i$. If 1 were in M, then 1 would lie in some L_j, giving $L_j \supseteq R1 = R$ and contradicting the supposition that L_j is proper. Thus, $1 \notin M$ and M is a proper left ideal. Suppose B is a left ideal properly containing M. Then B properly contains each $L_i (i \in I)$, and the collection of left ideals consisting of B and all $L_i (i \in I)$ is a nest of left ideals that properly contains the nest $\{L_i \mid i \in I\}$. The property of maximality possessed by $\{L_i \mid i \in I\}$ requires $B = R$. Thus, M is a maximal left ideal.

Now let M by *any* maximal left ideal of R. Let S be any submodule of R/M. Then $B = \{r \in R \mid r + M \in S\}$ is a left ideal of R satisfying $M \subseteq B \subseteq R$. By the maximality of M, we have either $M = B$ and S is the zero submodule of R/M or $B = R$ and $S = R/M$. Thus, R/M is simple.

It will be convenient to give the definition of a uniform module after defining the following very natural concept.

DEFINITION. An R-module is *indecomposable* if it is not the direct sum of any pair of proper submodules.

Simple modules are necessarily indecomposable. The concept of a uniform module is less restrictive than that of a simple module and more restrictive than that of an indecomposable module.

DEFINITION. An R-module is *uniform* if each of its submodules is indecomposable.

Stated another way, a module is uniform if and only if each pair of nonzero submodules has a nonzero intersection. The Z-module Q and each of its submodules are uniform. Any nonzero uniform Q-module must be isomorphic with Q.

Uniform modules have also been called *absolutely indecomposable modules*, which is a very descriptive label, especially in view of the following observations.

OBSERVATION 3. Let S be a uniform submodule of a direct sum $M \oplus M'$, and let p, p' be the associated projections onto M and M'. Then either the restriction of p to S or the restriction of p' to S is a monomorphism.

Verification: If neither of these projections were monic, we would have two nonzero submodules, $\ker(p) \cap S$ and $\ker(p') \cap S$, of S that would have intersection zero since $\ker(p) \cap \ker(p') = 0$.

The previous observation is not very surprising. On first contact, the next one is likely to be. Its verification will expose a feature of uniform modules that is a basic tool in their use. This feature may be informally described by saying that each nonzero element of a uniform module is necessarily interrelated with the overall structure of the module. The situation is a little weaker than in the case of simple modules, where a nonzero element must be a generator.

OBSERVATION 4. If S is a uniform submodule of $M = \oplus\{M_i \mid i \in I\}$, then, for some $j \in I$, the restriction to S of the jth projection p_j is a monomorphism.

Verification: If S is 0, then any j will do; so suppose $S \neq 0$, and choose $0 \neq s \in S$. Let J be the set of those $i \in I$ for which the ith component of s is not zero. Then $M = (\oplus\{M_i \mid i \in J\}) \oplus (\oplus\{M_i \mid i \in I \setminus J\})$. By the previous observation, the restriction to S of one of the two associated projections must be monic. It cannot be the projection onto the second summand since this projection sends s into 0. Thus, the projection of S into the first summand is monic. Let this projection be denoted by p, and let $S' = p(S)$. Let's shift our attention to S'. Now S' is a submodule of $\oplus\{M_i \mid i \in J\}$, and the importance of this reduction is in the fact that J is a finite set. A finite number of applications of Observation 3 yields a $j \in J$ for which the projection of S' into M_j is monic. (Exercise 10 requests a formalization of this process into a finite-induction argument.) The composite projection of S onto S' into M_j is also monic, as required.

Direct Summands of the Module R. The summands of the R-module R are tied to certain special elements of R.

DEFINITION. An element e of a ring is an *idempotent* if $ee = e$.

OBSERVATION 5. A submodule S of the R-module R is a summand if and only if $S = Re$ for an idempotent e in R.

Verification: Suppose $R = S \oplus S'$ and $1 = e + e'$ is the associated decomposition of 1. Since e is in S and S is a left ideal, we have $Re \subseteq S$. Let s be any element of S. From $s = s1 = se + se'$, $se \in S$, $se' \in S'$ and the directness of $R = S \oplus S'$, we conclude that $s = se \in Re$. Thus, $S = Re$. Moreover, from the special case in which $s = e$, we have $e = ee$, and e is an idempotent. Thus, if a submodule of R is a summand, it is an idempotent-generated left ideal.

Suppose now that e is an idempotent in R. We will verify that $R = Re \oplus R(1 - e)$ is a direct sum decomposition of R. For each r in R, we have $r = re + r(1 - e)$, which shows that $R = Re + R(1 - e)$. But $Re \cap R(1 - e) = 0$ since, if $r'e = r''(1 - e)$ holds for some r', r'' in R, we have $r'e = r'ee = r''(1 - e)e = r''(e - ee) = 0$. Thus, each idempotent-generated left ideal of R is a summand of R.

Because of their importance in structure theory, modules that are isomorphic with summands of R have been given a special name.

DEFINITION. A *principal R-module* is an R-module that is isomorphic with a direct summand of the R-module R.

To indicate the significance of the classes of modules I have introduced thus far, I will indicate their function in our structural studies chapter by chapter. The fundamental building block used in Chapter 2 is the R-module R itself. For Chapter 3 the basic building blocks are simple modules. In Chapter 4 our blocks will be left ideals of R. The uniform cyclic R-modules are fundamental blocks used throughout Chapter 9; in the final section of Chapter 9 the building blocks are also principal modules. In § 2 of Chapter 10 the building blocks are again uniform cyclic modules. The finitely generated left ideals of R are the blocks used in §§ 3 and 4 of Chapter 11 (those in § 4 are principal modules). Chapter 13 returns to R itself as the basic building block.

Aside from some elaborate decoration that must be added in certain cases to uniform cyclic modules (to be explained in Chapter 7), we have all the basic building blocks that are used in the theorems of this book. Our construction tool will be the direct sum.

Virtually all authors use the term *semisimple module* to mean a module that is the direct sum of simple modules. I will expand this use of the prefix *semi* to a broad convention: *the prefix semi, when applied to a type of module (but not necessarily to a type of ring), will mean "the direct sum of."* Thus, for example, by a semiprincipal module we will mean a module that is a direct sum of principal modules. Semicyclic, semiuniform, and so on now have clear meanings. Our structure theorems will have the following form: each module of some specified type is *semi* some structurally more transparent type.

§ 3 USING THE IDENTITY OF R (PROJECTIVITY)

We have already used the identity of R to discuss cyclic modules. There is another elementary use of the identity that gives strong conclusions concerning mappings out of (and also onto) R. In discussing this use of the identity we will meet one of the basic themes of module theory —that is, projectivity.

OBSERVATION 6. Let $h : M \to R$ be an epimorphism. For any element c of M for which $h(c) = 1$, the restriction of h to the cyclic submodule Rc is an isomorphism of Rc onto R. Further, $M = Rc \oplus \ker(h)$.

Verification: The computation $h(rc) = rh(c) = r1 = r$, which is valid for every r in R, shows that the restriction of h to Rc is bijective and consequently that $Rc \cap \ker(h) = 0$. Thus, we need only verify that $M = Rc + \ker(h)$. Let m be any element of M, and let $r = h(m)$. Then $m = rc + (m - rc)$, $rc \in Rc$, and $(m - rc) \in \ker(h)$, since $h(m - rc) = h(m) - rh(c) = r - r1 = 0$.

Observation 6 can be derived from the following more sweeping observation.

OBSERVATION 7. For any epimorphism $h : A \to B$ and any map $k : R \to B$, there is a map $k' : R \to A$ for which $k = hk'$.

Verification: We need only concern ourselves with the element 1 of R. Let c be any element of A for which $h(c) = k(1)$. There is a unique map $k' : R \to A$ for which $k'(1) = c$. Since $hk'(1) = h(c) = k(1)$ (that is, since k and hk' agree on the identity), $k = hk'$.

The map k' is called a lifting of k or, in more detail, a lifting of k through h. In this language the observation may be stated concisely: *maps with domain R can be lifted through epimorphisms.* This map-lifting property has proved to be so significant that it has been incorporated into the following definition.

DEFINITION. An R-module P is *projective* if, for each epimorphism $h : A \to B$ and each map $k : P \to B$, there is a map $k' : P \to A$ for which $k = hk'$.

We have shown that R is projective. Which quotient modules of R are projective?

OBSERVATION 8. A cyclic module is projective if and only if it is a principal module.

Verification: For any cyclic R-module we have an epimorphism $h : R \to C$ (given by $h(r) = rc$, where c is a generator of C). Suppose C is projective. Let $k : C \to C$ be the identity map. There is a $k' : C \to R$ for which $k = hk'$. From Chapter 0, k' is a splitting map for h (that is, $R = \ker(h) \oplus \operatorname{im}(k')$); therefore, $C \cong \operatorname{im}(k')$ is a principal module.

Now suppose that C is a principal module. Then C is isomorphic with a direct summand of R, and we may as well assume $R = C \oplus D$.

Let $h : A \to B$ be any epimorphism, and let $k : C \to B$ be any map. Define $g : R \to B$ by $g(c + d) = k(c)$. Since g is a map and R is projective, there is a map $g' : R \to A$ for which $g = hg'$. Let k' be the restriction of g' to C. Since the restriction of g to C is k, we have $k = hk'$ as required.

Rings with the property that *all* of their left ideals are projective will arise in the course of our investigations in Chapter 4.

EXERCISES

1. Regard the ring Z of integers as a module over itself and study Z as follows:
   a. Determine the submodules of Z.
   b. Determine the quotient modules of Z.
   c. Which submodules of Z are maximal?
   d. Determine the simple Z-modules.
   e. Which cyclic Z-modules are indecomposable?
   f. Which cyclic Z-modules are uniform?
   g. Which elements of Z are idempotent?
   h. Determine the principal Z-modules.
   i. Which cyclic Z-modules are semisimple?
   j. Which cyclic Z-modules are semiuniform?
2. Regard the ring Q of rational numbers as a module over itself. Carry out Exercise 1, replacing Z with Q.
3. For some third ring R of your own choice, regard R as a module over itself and attempt to carry out Exercise 1, replacing Z with R. Here are some choices of R that you may wish to consider: (1) the ring Z_n of integers reduced modulo n for some integer n, (2) the ring $Q[x]$ of all polynomials over Q in the indeterminant x, and (3) the ring $Q^{2 \times 2}$ of all 2×2 matrices over Q.
4. Let $h : A \to Q$ be a Q-module homomorphism for which $\text{im}(h) \neq 0$.
   a. Is h necessarily surjective?
   b. Show that A contains a summand isomorphic to Q.
5. Let $h : A \to Z$ be a Z-module homomorphism for which $\text{im}(h) \neq 0$.
   a. Is h necessarily surjective?
   b. Show that A contains a summand isomorphic to Z.
6. Let $h : A \to Z_4$ be a Z_4-module homomorphism for which $\text{im}(h) \neq 0$. Can you necessarily conclude that A contains a summand isomorphic to Z_4?
7. Prove that for each proper left ideal A of a ring R there is a maximal left ideal B of R containing A.
8. Let S be a simple R-module. Prove that R contains a maximal left ideal M for which R/M is isomorphic with S.

9. a. Show that there exists a projective simple R-module if and only if some maximal left ideal of R is a direct summand of R.
   b. Show that every simple R-module is projective if and only if every maximal left ideal of R is a direct summand of R.

10. Prove the following assertion using Observation 3 and finite induction on n. If S is a uniform submodule of a direct sum $M = M_1 \oplus \cdots \oplus M_n$, then, for at least one of the projections $p_i : M \to M_i$, the restriction of p_i to S is a monomorphism.

11. Let R be an integral domain. Regard R as a module over itself. Show that R is a uniform R-module.

12. Let A be an R-module that is not uniform. Construct modules M and M' such that $M \oplus M'$ contains a submodule S that is isomorphic with A and for which neither the projection of S into M nor the projection of S into M' is a monomorphism.

13. Let M be an R-module and let $\{A_i \mid i \in I\}$ be a family of submodules of M that satisfies the following condition: $0 = \Sigma\{a_i \mid i \in F\}$ can hold for a finite subset F of I and elements $a_i \in A_i$ $(i \in F)$ only when $a_i = 0$ $(i \in F)$. Show that for the submodule S of M generated by $\cup\{A_i \mid i \in I\}$ we have $S = \oplus\{A_i \mid i \in I\}$. (This fact is used in the verification of Observation 5 of Chapter 2.)

PART ONE

PROJECTIVITY

CHAPTER 2

DIRECT SUMS OF COPIES OF R (FREE MODULES)

§ 1 INTRODUCTION

In Chapter 1 we examined the ring R regarded as a module over itself. We made an introductory survey of the summands, submodules, and quotient modules of R. Through the use of the identity element of R, we encountered the concept of a projective module. The pattern of development of Chapter 1 and the concepts introduced there provide an introduction to Part One. The role played by the R-module R in Chapter 1 will be taken over by the class of all direct sums of copies of the module R or, more generally, by the free R-modules.

DEFINITION. An R-module is *free* if it is isomorphic to a direct sum of copies of the module R.

Chapter 2 is concerned with the free modules themselves. The remaining three chapters of Part One take up successively, the direct summands of free modules, the submodules of free modules, and the quotient modules of free modules. In the course of extending our attention from free modules to these broader classes, we obtain a thorough introduction to the concept of projectivity. We will return to deeper structural investigations of projective modules in Part Two, Chapter 9, § 5, and again in the final three chapters of Part Three.

§ 2 BASES

The definition I have chosen for the concept of a free module is itself a structural description of these modules. However, since free modules play such a basic role in module theory, I will reformulate this definition in terms of sets of elements, and we will practice using this alternate, element-wise description.

For an arbitrary ring R and an arbitrary set I, let $\{R_i \mid i \in I\}$ be a family of copies of the R-module R indexed by the set I. Thus, $R_i = R$ for each i in I. Let $F = \oplus\{R_i \mid i \in I\}$. Then the set F consists of those functions $f : I \to R$ for which $f(i) = 0$ for all but a finite number of i in I. With each i in I associate the element b_i of F defined by the conditions $b_i(j) = 1$ if $j = i$ and $b_i(j) = 0$ if $j \neq i$. Let $B = \{b_i \mid i \in I\}$. Let f be an arbitrary element of F. For each i in I, let $r_i = f(i)$. Then $r_i = 0$ for all but a finite number of i in I, and consequently the sum $\Sigma\{r_i b_i \mid i \in I\}$ is defined and is an element of F. Since, for each j in I, $(\Sigma r_i b_i)(j) = \Sigma r_i b_i(j) = r_j$, we see that $f = \Sigma r_i b_i$. Thus, B is a set of generators of F. But B has a further property. Suppose that $\{r'_i \mid i \in I\}$ is any indexed set of elements of R for which $r'_i = 0$ for all but a finite number of i in I and that $f = \Sigma r'_i b_i$. Then for each j in I we have $f(j) = (\Sigma r'_i b_i)(j) = \Sigma r'_i b_i(j) = r'_j$, and consequently $r'_j = r_j$ for every j in I. This uniqueness of the representation $f = \Sigma r_i b_i$ shows that B is a basis of F in the following sense.

DEFINITION. A subset B of an R-module A is a *basis* of A if for each element a in A there is a unique indexed set $\{r_b \mid b \in B\}$ of elements of R for which $r_b = 0$ for all but a finite number of b in B and $a = \Sigma\{r_b b \mid b \in B\}$.

Not all modules possess bases. The possession of a basis is invariant under isomorphism. If $h : A \to A'$ is an isomorphism of R-modules and B is a basis of A, then $h(B) = \{h(b) \mid b \in B\}$ is a basis of A' (see Exercise 2). It follows that every free module has a basis. We will show that the possession of a basis characterizes the free modules. For this purpose, the following powerful map-building tool will be convenient.

LEMMA. Let A be an R-module that has a basis B, and let M be an arbitrary R-module. For each function $g : B \to M$ there is a unique map $h : A \to M$ for which $h(b) = g(b)$ for every b in B.

Proof: For each $a \in A$ we have a unique representation $a = \Sigma\{r_b b \mid b \in B, r_b \in R\}$. *If there is a map* h that agrees with g on B, it must satisfy $h(a) = h(\Sigma r_b b) = \Sigma r_b h(b) = \Sigma r_b g(b)$. This shows that there is *at most* one map of A into M that agrees with g on B. It also tells us that to complete the proof of the lemma we need only define the function $h : A \to M$ by $h(a) = \Sigma r_b g(b)$ and verify that it is a map and that it agrees with g on B. For $a = \Sigma r_b b \in A$ and $r \in R$ we have $h(ra) = h(\Sigma r r_b b)$ $= \Sigma r r_b g(b) = r\Sigma r_b g(b) = rh(a)$. For $a = \Sigma r_b b$ and $a' = \Sigma r'_b b$ we have $h(a + a') = h(\Sigma(r_b + r'_b)b) = \Sigma(r_b + r'_b)g(b) = \Sigma r_b g(b) + \Sigma r'_b g(b) = h(a) + h(a')$. Finally, for an element $b' \in B$ the representation $b' = \Sigma r_b b$ is simply the one given by $r_{b'} = 1$ and $r_b = 0$ if $b \neq b'$. Thus, $h(b') = \Sigma r_b g(b) = r_{b'} g(b')$ $= g(b')$.

From the uniqueness condition in the lemma we see that, if A is a module with a basis B, to verify that maps $h : A \to M$ and $k : A \to M$ are identical, we need only show that they agree on B. It is now convenient to give the following basic characterization of free modules.

OBSERVATION 1. A module is free if and only if it contains a basis.

Verification: We have already seen that a free module must have a basis; suppose that A is an R-module with a basis I. Use I to index a family $\{R_i \mid i \in I\}$ of copies of R and form $\oplus\{R_i \mid i \in I\}$. Let $B = \{b_i \mid i \in I\}$ be the basis of $\oplus R_i$ as constructed in the initial discussion of this section. Let $g : I \to B$ be the bijection given by $g(i) = b_i$, and let g' be the inverse of g. There is a unique map $h : A \to \oplus R_i$ that agrees with g on I, and there is a unique map $h' : \oplus R_i \to A$ that agrees with g' on B. Then $h'h$ is a map from A into A that agrees with the identity function, $g'g$, on I, and hh' is a map from $\oplus R_i$ into $\oplus R_i$ that agrees with the identity function, gg', on B. It follows that $h'h$ and hh' are identity maps and that h and h' are a pair of inverse isomorphisms.

This observation suggests that a routine way of showing that a module is free is to construct a basis of the module.

A free module may possess more than one basis. For the Z-module Z, $\{1\}$ is a basis and so is $\{-1\}$. For $Z \oplus Z$, the pair of elements $\{(1,0), (n,1)\}$ is a basis for any n in Z. Must two bases of the same free module have the same cardinal number? The answer may be surprising. If the module has an infinite basis, order prevails. *If F is a free module that has an infinite basis, then any two bases of F must have the same cardinal.* This fact is a consequence of the following observation, which contains extra information.

OBSERVATION 2. If F is a free R-module having a basis B of infinite cardinal β, then each subset A of F having cardinal $\alpha < \beta$ is contained in a proper direct summand of F.

Verification: Each a in A has a unique representation $a = \Sigma r_b b$ with $b \in B$ and $r_b \in R$. Let $B(a) = \{b \in B \mid r_b \neq 0\}$. Then each $B(a)$ is finite and the cardinal number γ of the union $C = \cup\{B(a) \mid a \in A\}$ is finite if A is finite and is equal to α if A is infinite. In either case, we have $\gamma < \beta$, $F = (\oplus\{Rb \mid b \in C\}) \oplus (\oplus\{Rb \mid b \in B \setminus C\})$, and $A \subseteq \oplus\{Rb \mid b \in C\} \neq F$.

If a free module does not have an infinite basis, chaos may reign. Among the exercises, the following information is developed. *There exists a ring R for which for every positive integer n there is a basis of the R-module R that has precisely n elements.* At later points in this book, we will show that for rings of particular types, any two bases of the same free module must have the same cardinal number.

I will make two final observations that give fundamental properties of free modules. They will be demonstrated through the use of the basis concept and its attendant lemma. They illustrate further that a basis is to a free R-module what the identity element is to the module R.

OBSERVATION 3. Free modules are projective.

Verification: Let F be a free module over a ring R. Let $h : A \to B$ be an arbitrary epimorphism of R-modules. Let $k : F \to B$ be a map. We must construct a map $k' : F \to A$ for which $hk' = k$. We begin by choosing a basis I of F. Since h is surjective, for each $i \in I$ there is an $a_i \in I$ for which $h(a_i) = k(i)$. Define a function $g : I \to A$ by $g(i) = a_i$. There is a unique map $k' : F \to A$ that agrees with g on I. For each $i \in I$, $hk'(i) = h(k'(i)) = h(g(i)) = h(a_i) = k(i)$. Thus, hk' and k agree on a basis of F, and we may conclude that $hk' = k$ as required.

OBSERVATION 4. If $h : A \to P$ is an epimorphism and P is projective, then h splits.

Verification: To show that h splits, we must show that there exists a map $k' : P \to A$ for which hk' is the identity map on P. If we let $k : P \to P$ be the identity map on P, then such a k' must exist by the projectivity of P.

It must be apparent that the specification of a basis of a free R-module F is very closely related to the specification of an isomorphism $F \cong \oplus\{R_i \mid i \in I\}$ of F with the direct sum of an indexed family of copies of R. I will close the discussion of bases by spelling out this relatonship. Suppose $B = \{b_i \mid i \in I\}$ is a basis of F. Then $F = \oplus\{Rb_i \mid i \in I\}$, and for each i in I the map $h_i : R \to Rb_i$ defined by $h_i(r) = rb_i$ is an isomorphism. The family of isomorphisms $\{h_i \mid i \in I\}$ may be assembled into an isomorphism $h : \oplus\{R_i \mid i \in I\} \to \oplus\{Rb_i \mid i \in I\} = F$, where $\{R_i \mid i \in I\}$ is a family of copies of R indexed by I. Conversely, let $h : \oplus\{R_i \mid i \in I\} \to F$ be an isomorphism, where $\{R_i \mid i \in I\}$ is an indexed family of copies of R. Let $B = \{b_i \mid i \in I\}$ be the basis of $\oplus R_i$ constructed in the initial discussion of this section. Then $\{h(b_i) \mid i \in I\}$ is a basis of F.

It is now clear that in discussing the bases of a free R-module F we have been discussing the various ways of forming isomorphisms $F \cong \oplus\{R_i \mid i \in I\}$ of F with direct sums of copies of R. We have learned of the existence of a ring R that has, for each positive integer n, a basis consisting of n elements. Expressing this fact in terms of direct sum representations gives, for such a ring, the following disconcerting sequence of R-module isomorphisms: $R \cong R \oplus R \cong R \oplus R \oplus R \cong \ldots.$

§ 3 INDEPENDENT SETS

This section is devoted to a concept of independence that provides a fundamental tool used in each part of this book.

DEFINITION. A subset S of an R-module A is *independent* if S does not contain 0 and the submodule of A generated by S is the internal direct sum of the cyclic submodules Rs generated by the elements s in S.

We will modify the word "independent" with the adverbs "uniformly" and "simply," producing the following meanings: The subset S is *uniformly independent* if it is independent and Rs is uniform for each s in S. It is *simply independent* if it is independent and Rs is simple for each s in S.

OBSERVATION 5. (1) The union of a nest of independent subsets of a module is independent. (2) If the members of the nest are uniformly independent, then the union is uniformly independent. (3) If the members of the nest are simply independent, then the union is simply independent.

Verification: Let $\{S_i \mid i \in I\}$ be a nest of independent subsets of an R-module A. Let $S = \cup\{S_i \mid i \in I\}$ and let B be the submodule generated by S. We must show that B is the internal direct sum of the submodules $\{Rs \mid s \in S\}$. To do this, we let F be any finite subset of S and $\{r_s \mid s \in F\}$ any family of elements of R for which $\Sigma\{r_s s \mid s \in F\} = 0$. We must show that $r_s s = 0$ for each $s \in F$. Since F is a *finite* subset of S and S is the union of the *nest* of sets $\{S_i \mid i \in I\}$, there is a j in I for which $F \subseteq S_j$. The independence of S_j then provides the required equalities: $r_s s = 0$ for each $s \in F$. The verification of (1) is complete; (2) and (3) follow immediately.

The concept of independence will always be employed by means of one of the three assertions of the following lemma.

LEMMA. Every R-module A contains (1) a maximal independent set, (2) a maximal uniformly independent set, and (3) a maximal simply independent set.

Proof: (1) Consider the collection of all those subsets of A that are independent. This collection is not empty, since the empty set is independent. The collection contains at least one maximal nest. Let $\{S_i \mid i \in I\}$ be such a maximal nest, and let $S = \cup\{S_i \mid i \in I\}$. Then S is independent and is not properly contained in any other independent subset. If S' were an independent subset of A, which properly contained S, then S' would properly contain each $S_i (i \in I)$; consequently, the collection of subsets consisting of S' and each $S_i (i \in I)$ would be a nest of independent subsets of A that properly contained the maximal nest $\{S_i \mid i \in I\}$. We have shown that S is a maximal independent subset of A.

(2) Reread the proof of (1) inserting the adverb "uniformly" before each occurrence of the word "independent." (3) Insert "simply" before each occurrence of "independent" in the proof of (1).

A basis of a free R-module is a maximal independent set. A basis of a nonzero free R-module is uniformly independent if and only if R is uniform; it is simply independent if and only if R is simple. Regard Z as a Z-module. The maximal independent subsets of Z are the singleton subsets $\{n\}$, where $n \neq 0$. The only maximal independent subsets of Z that are bases are $\{1\}$ and $\{-1\}$. Each independent subset of Z is uniformly independent. The only simply independent subset of Z is the empty set that generates the submodule $\{0\}$.

§ 4 DIVISION RINGS

We have developed some familiarity with free modules. If a ring R has the property that all its modules are free, then we have some familiarity with all R-modules. It is therefore natural to ask "Which rings have the property that all their modules are free?" The answer to this question will pick out the following class of rings.

DEFINITION. A ring R is a *division ring* if for each pair of elements a, b in R with $a \neq 0$ there is an element x in R for which $xa = b$.

Modules over a division ring R are often called *vector spaces over* R or R-*spaces*. A commutative division ring is called a *field*. The classical example of a division ring that is not a field is given by the quaternions that were studied extensively by W. R. Hamilton and are described in Exersice 7. A suprising, but classical, result of Wedderburn asserts that every finite division ring is a field (see § 2, Chapter 7 of Herstein [1964]).

If R is a division ring, then R is simple when it is regarded as an R-module: Let L be any nonzero left ideal of R, and let a be a nonzero element of L. For any b in R there is an x in R for which $xa = b$, and consequently b must be in L. Thus, $L = R$, confirming that R is simple. Since R is simple, every nonzero cyclic R-module is isomorphic to R and is therefore simple and also uniform. Thus, for modules over division rings, every independent subset is uniformly independent and also simply independent.

PROPOSITION 1. A ring R has the property that all its modules are free if and only if R is a division ring.

Proof: Let R be a ring for which every R-module is free. Let S be a simple R-module. Since S is free, it is isomorphic with a direct sum of copies of R. Since S is simple, it is indecomposable, and therefore S is isomorphic with R. Thus, R itself is simple as an R-module, and the only nonzero left ideal of R is R itself. For any nonzero element a in R, Ra is a nonzero left ideal, and consequently $Ra = R$. The equality guarantees that for each b in R there is an x in R for which $xa = b$. Thus, R is a division ring.

Suppose now that R is a division ring and that A is an arbitrary R-module. Let S be a maximal independent subset of A. Since, for each s in S, Rs is a nonzero cyclic module, we have $Rs \cong R$. Thus, to show that A is free we need only show that S generates A. For a in $A \setminus S$, $S \cup \{a\}$ is not independent since S is a maximal independent subset of A. Consequently, there are elements r, r_s in R, not all of which are zero, such that $ra + \Sigma\{r_s s \mid s \in S\} = 0$. Since S is independent, $r \neq 0$. Since R is a division ring, there is an element x in R for which $xr = 1$. Then $a = xra = x(-\Sigma r_s s) = \Sigma(-xr_s)s$. Thus, S generates A, and A is free.

OBSERVATION 6. Two bases of the same (free) module A over a division ring R must have the same cardinal number.

Verification: By Observation 2 we need only consider R-modules that do not have an infinite basis. If A has the empty set as a basis, then $A = 0$ and A has no other basis. This gives us a foundation for a finite-induction argument. Let A have a basis $a_1, \ldots, a_n$, and suppose that the observation holds for all modules having a basis of fewer than n elements. We need only assume that $b_1, \ldots, b_{n+k}$ is a basis of A for which k is a nonnegative integer and prove that $k = 0$. Let $b_1 = r_1 a_1 + \cdots + r_n a_n$. There is a j for which $r_j \neq 0$. For notational convenience, reindex the basis a_i so that $r_1 \neq 0$. Then $b_1, a_2, \ldots, a_n$ is also a basis of A (see Exercise 11). Let B be the submodule generated by $a_2, \ldots, a_n$, and let C be the submodule generated by $b_2, \ldots, b_{n+k}$. Then $A = Rb_1 \oplus B = Rb_1 \oplus C$, and consequently $B \cong A/Rb_1 \cong C$. Then B has a basis of $n - 1$ elements and also a basis of $n + k - 1$ elements. We conclude that $n - 1 = n + k - 1$ and $k = 0$ as required.

The scope of Observation 6 is greatly extended by Observation 7.

OBSERVATION 7. Let R be a ring that is a homomorphic image of a ring R'. If R has the property that any two bases of the same free R-module have the same cardinal number, then R' also has this property.

The steps in the verification of this observation are outlined in Exercises 13 and 14. A widely used consequence of Observation 7 for commutative rings appears in Exercise 15. For a guide to further results related to Observations 6 and 7, see page 41 of Cohn [1971].

EXERCISES

1. Is a trivial (zero) R-module free? If so, what is its basis?

2. Let $h : A \to A'$ be an isomorphism of the R-module A onto the R-module A', and suppose A has a basis B. Show that $h(B) = \{h(b) \mid b \in B\}$ is a basis of A'.

3. Let R be an arbitrary ring. We will construct from R a new ring, R', that will be an infinite dimensional analogue of the ring of n by n matrices over R. The elements of R' will be the so-called row-finite matrices: $\{(r_{ij}) \mid i, j$ are positive integers; $r_{ij} \in R$; for each i, $r_{ij} = 0$ for all but finitely many $j\}$. The addition and multiplication in R' are entirely analogous to the usual operations with $n \times n$ matrices: $(r_{ij}) + (r'_{ij}) = (s_{ij})$, where $s_{ij} = r_{ij} + r'_{ij}$; and $(r_{ij}) \cdot (r'_{ij}) = (\mu_{ij})$, where $\mu_{ij} = \Sigma\{r_{ik}r'_{kj} \mid k$ is a positive integer$\}$. Notice that the apparently infinite sum is actually defined because r_{ik} is nonzero for only finitely many k.
   a. Verify that R' is a ring.
   b. Let b_1 be the matrix (r_{ij}) for which $r_{ij} = 1$ when $j = 2(i - 1) + 1$ and $r_{ij} = 0$ otherwise. Let b_2 be the matrix (r_{ij}) for which $r_{ij} = 1$ when $j = 2(i - 1) + 2$ and $r_{ij} = 0$ otherwise. Show that $\{b_1, b_2\}$ is a basis of R'.
   c. For each positive integer n, specify a basis of R' that consists of n elements.
   d. Does R' have an infinite basis?

4. Let Z_6 be the additive group of integers modulo six. Regard Z_6 as a Z-module.
   a. Find a uniformly independent subset of Z_6 that generates Z_6. Is this set simply independent?
   b. Find an independent subset of Z_6 that is not uniformly independent. Is this set a basis of the module?

5. The ring Z_6 may be regarded as a module over itself. Rework the previous exercise regarding Z_6 as a Z_6-module rather than a Z-module.

6. Regard the ring Z_{12} as a module over itself.
   a. Does a uniformly independent subset exist that generates this module?
   b. Does every maximal (uniformly) independent subset generate this module?
   c. Is there a simply independent subset that generates this module?

7. Let H be the set of formal expressions $\{a + bi + cj + dk \mid a, b, c, d$ are rational numbers$\}$. Define addition in H by $(a + bi + cj + dk) + (a' + b'i + c'j + d'k) = (a + a') + (b + b')i + (c + c')j + (d + d')k$. Define multiplication in H by $(a + bi + cj + dk) \cdot (a' + b'i + c'j + d'k) = (aa' - bb' - cc' - dd') + (ab' + ba' + cd' - dc')i + (ac' + ca' + db' - bd')j + (ad' + da' + bc' - cb')k$.

Verify that H is a division ring that is not commutative. This division ring is called the ring of *rational quaternions*. Reread your verification concerning H but with the understanding that the coefficients a, b, c, d, and so on are allowed to be arbitrary real numbers. This gives a second division ring called the ring of *real quaternions*.

8. Show that a ring R is a division ring if and only if $R\setminus\{0\}$ is a group under multiplication.

9. Show that a ring R is a division ring if and only if for each pair of elements a, b in R with $a \neq 0$ there is an element y in R for which $ay = b$.

10. For which rings R is the following assertion correct? Each maximal independent subset of each R-module A is necessarily a basis of A.

11. Let R be a division ring, and let A be an R-module that possesses a basis $a_1, \ldots, a_n$. Show that, if $b = r_1a_1 + \cdots + r_na_n$ and $r_1 \neq 0$, then $b, a_2, \ldots, a_n$ is also a basis of A. You may wish to generalize this observation.

12. Let A be an R-module. An R-module homomorphism of A into itself is called an *endomorphism* of the R-module A. Let E be the set of all endomorphisms of A.
    a. Show that the sum $f + g$ of two endomorphisms f, g of A is again an endomorphism and also that the composite $f \circ g$ is an endomorphism. The definitions are as usual: $(f + g)(a) = f(a) + g(a)$, and $f \circ g(a) = f(g(a))$.
    b. Verify that E is a ring where addition is given by $f + g$ and multiplication is given by composition $f \cdot g = f \circ g$.
    c. Prove that, if A is a simple R-module, then E is a division ring.
    d. Let R' be an arbitrary division ring. Can you find a ring R having a simple R-module A for which R' is isomorphic to the endomorphism ring of A?
    e. List the simple Z-modules and their endomorphism rings.
    f. Let p be a prime and let $Z_p[x]$ be the ring of polynomials in x with coefficients in the field Z_p of integers mod p. Describe the simple $Z_p[x]$-modules and their endomorphism rings. Is every finite field of characteristic p isomorphic to one of these endomorphism rings?

13. Let R' be a ring, and let J be a proper (two-sided) ideal of R'. Let R be the ring R'/J. For each R'-module A, let JA be the R'-submodule generated by the subset $\{ja \mid j \in J, a \in A\}$.
    a. Verify that the R'-module A/JA is an R-module with respect to the operation defined, for all $r \in R'$ and $a \in A$, by $(r + J) \cdot (a + JA) = ra + JA$.
    b. Show that, if A is a free R'-module with basis $b_1, \ldots, b_n$, then A/JA is a free R-module with basis $b_1 + JA, \ldots, b_n + JA$.

14. Prove Observation 7.

15. Let R be a commutative ring, and let M be a maximal left ideal of R.
    a. Observe that M is a two-sided ideal, that R/M is a field, and that the natural mapping of R onto R/M is a ring homomorphism.
    b. Prove that any two bases of the same free module over R must have the same cardinal number.

CHAPTER 3

SUMMANDS OF FREE MODULES (PROJECTIVE MODULES)

§ 1 INTRODUCTION

In this chapter we expand our attention from the class of free modules to the class of all direct summands of free modules. Our goal is to develop as much insight into the structure of these modules as possible. Only the first steps can be taken in this chapter. I will give a survey discussion of this problem after making an observation that will allow us to shorten the phrase "a module isomorphic to a direct summand of a free module" to a single familiar word.

OBSERVATION 1. An R-module is isomorphic to a direct summand of a free R-module if and only if it is projective.

Verification: Let $F = C \oplus D$, where F is a free R-module. Let $h : A \to B$ be an epimorphism of R-modules, and let $k : C \to B$ be a map.

We must construct a map $k' : C \to A$ for which $hk' = k$. To do this, we first define the map $g : F \to B$ by $g(c + d) = k(c)$, where c and d are in C and D respectively; notice that k is the restriction of g to C. Since F is projective, there is a map $g' : F \to A$ for which $hg' = g$. We may now define k' to be the restriction of g' to C. Then $hk' = k$ follows from $hg' = g$ by restricting domains to C.

Now let P be a projective R-module. Ignore (for a moment) the algebraic structure on P, and regard P only as a set. Use P to index a family $\{R_p \mid p \in P\}$ of copies of R, and form the free R-module $F = \oplus\{R_p \mid p \in P\}$. As a basis for F we have $B = \{b_p \mid p \in P\}$, where, for p and q in P, $b_p(q) = 1$ if $q = p$ and $b_p(q) = 0$ if $q \neq p$. The function $f : B \to P$ defined by $f(b_p) = p$ extends uniquely to an R-homomorphism $h : F \to P$. This map is an epimorphism since f itself is surjective. Since P is projective, we know by Observation 4 of Chapter 2 that h splits, and consequently $F = \ker(h) \oplus P'$ for a submodule P' isomorphic with P.

This observation has two immediate consequences, which will be used frequently and without reference.

COROLLARY 1. A direct summand of a projective module is projective.

COROLLARY 2. The direct sum of any family of projective modules is projective.

In asking for structural insight into direct summands of free modules, we have initiated the investigation of the structure of projective modules. Let us make a survey of all the projective R-modules that we might regard as immediately available from the R-module R. Of course R itself is the simplest example, but in Chapter 1 we also examined the submodules, quotient modules, and summands of R. Under what circumstances were these associated modules projective? A submodule of R is simply a left ideal, which may or may not be projective. A cyclic R-module is projective if and only if it is isomorphic with a summand of R. Finally, every summand of R is a projective left ideal. Our search for projectives among the modules associated with R has yielded only the projective left ideals of R. When we close this class under direct sums, we have what we consider to be the broadest class of examples of projective modules that are readily available from R itself—that is, *the direct sums of projective left ideals of R*.

If we hope for very strong restrictions on the structure of the projective modules over a given ring R, then an optimistic wish would be for a theorem of the following form. For the ring R, an R-module is projective if and only if it is isomorphic to a direct sum of projective left ideals of R. Each of the three basic structure theorems for projective modules that will be given in this book justifies this optimistic wish for a particular class of rings—quasi-Frobenius rings (Chapter 9), semihereditary rings (Chapter 13), and local rings (Chapter 13). These structure theorems appear late in the book after the consideration of more basic topics. The structural results presented in this chapter arise from an investigation of modules over rings R for which *every* R-module is projective.

§ 2 SEMISIMPLE RINGS

For which rings R is every R-module projective? I will show that the answer is the class of rings described as follows.

DEFINITION. A ring R is a *semisimple ring* if R is semisimple when regarded as an R-module.

According to this definition, R is semisimple if $R = \oplus\{S_i \mid i \in I\}$, where each S_i is a simple left ideal. An elementary, but important, fact is that I is necessarily finite. Let $1 = \Sigma\{s_i \mid i \in I\}$ be the representation for the identity element with respect to this direct decomposition of R. Then $s_i \neq 0$ for only a finite subset F of I. Thus, $1 = \Sigma\{s_i \mid i \in F\}$, and for each r in R we have $r = r1 = r\Sigma\{s_i \mid i \in F\} = \Sigma\{rs_i \mid i \in F\} \in \oplus\{S_i \mid i \in F\}$. It follows that $R = \oplus\{S_i \mid i \in F\}$ and $I = F$. For the remainder of our discussion, when we express a semisimple R in the form $R = \oplus\{S_i \mid i \in F\}$, it will be assumed that the S_i are simple left ideals. (Notice that the semisimple rings for which the index set F is a singleton are precisely the division rings.)

The key to the structure of modules over semisimple rings is the following observation.

OBSERVATION 2. For each left ideal L of a semisimple ring $R = \oplus\{S_i \mid i = 1, \ldots, n\}$ there is a subset F of the index set for which $R = L \oplus (\oplus\{S_i \mid i \in F\})$.

Verification: Define L_1 to be $L \oplus S_1$ if $L \cap S_1 = 0$ and to be L if $L \cap S_1 = S_1$. Continue in this way for each subscript. Define L_j to be $L_{j-1} \oplus S_j$ if $L_{j-1} \cap S_j = 0$ and to be L_{j-1} if $L_{j-1} \cap S_j = S_j$. Then

$L_n = L \oplus (\oplus\{S_i | i \in F\})$ for some subset F of $\{1, \ldots, n\}$, and L_n contains each S_j $(1 \le j \le n)$. Consequently, $R = L_n = L \oplus (\oplus\{S_i | i \in F\})$.

COROLLARY 3. The following conditions hold for a semisimple ring $R = \oplus\{S_i \mid i = 1, \ldots, n\}$: (1) each left ideal is cyclic; (2) each cyclic module is semisimple; (3) each nonzero uniform cyclic module is simple; (4) each simple module is isomorphic to S_j for some j $(1 \le j \le n)$; and (5) each semisimple module is projective.

By using maximal simply independent sets, we can now determine the structure of all modules over a semisimple ring. Notice that for modules over a semisimple ring R every uniformly independent set is simply independent; but, unless R is a division ring, there are independent sets that are not simply independent.

THEOREM 1. Every module over a semisimple ring is semisimple.

Proof: Let $R = \oplus\{S_i \mid 1 \le i \le n\}$ be a semisimple ring, and let A be an arbitrary R-module. Let M be a maximal simply independent subset of A, and let B be the semisimple submodule $\oplus\{Rm \mid m \in M\}$. By the maximality of M, B must contain every simple submodule of A. Then B also contains every semisimple submodule of A, and, since every cyclic R-module is semisimple, B contains every element of A. Thus, $A = B$ is semisimple.

When a semisimple ring R is expressed in the form $R = \oplus\{S_i \mid i = 1, \ldots, n\}$, every simple R-module is isomorphic to one of the S_j. This suggests that we rearrange the S_i and reindex them with double subscripts so that we produce a representation $R = (S_{11} \oplus \cdots \oplus S_{1k(1)}) \oplus (S_{21} \oplus \cdots \oplus S_{2k(2)}) \oplus \cdots \oplus (S_{m1} \oplus \cdots \oplus S_{mk(m)})$, where S_{ij} is isomorphic with $S_{i'j'}$ if and only if $i = i'$. With this indexing we see that every simple R-module is isomorphic with precisely one member of the list: $S_{11}, S_{21}, \ldots, S_{m1}$. With any m-tuple $(\kappa_1, \ldots, \kappa_m)$ of cardinal numbers, associate the R-module that is the direct sum of κ_1 copies of S_{11}, κ_2 copies of $S_{21}, \ldots,$ and κ_m copies of S_{m1}. By our structure theorem (and Exercise 13), each R-module is isomorphic with the module associated with precisely one m-tuple of cardinal numbers.

We can now answer very conveniently the initial question of this section: for which rings R is every R-module projective?

PROPOSITION 2. A ring R has the property that all its (simple) modules are projective if and only if R is semisimple.

Proof: If R is semisimple, then each of its modules is semisimple by Theorem 1 and projective by Corollary 3. Suppose now that R is a ring with the property that each of its simple modules is projective. Let M be a maximal simply independent subset of R. Then $\oplus\{Rm \mid m \in M\}$ is a semisimple submodule of R which, by the maximality condition on M, contains every simple submodule of R. Let $A = \oplus\{Rm \mid m \in M\}$. I will show that R is semisimple by showing that $R = A$. Suppose that $A \neq R$. By Exercise 7 of Chapter 1, there is a maximal left ideal B of R that contains A. Since R/B is simple, it must be projective. By Observation 4 of Chapter 2, $R = B \oplus S$ for a submodule $S \cong R/B$ of R. Since S is a simple submodule of R, S is contained in B, contradicting the directness of the decomposition $R = B \oplus S$. From the contradiction we conclude that $A = R$.

The two remaining sections of Chapter 3 are discussions of examples of semisimple rings.

§ 3 MATRIX RINGS

We will discuss in an informal manner some examples of rings of matrices that are semisimple. Let D be a division ring. Let $R = D^{2\times 2}$ be the ring of all 2×2 matrices with entries from D. Let

$$S_1 = \left\{\begin{pmatrix} a & 0 \\ b & 0 \end{pmatrix} \middle| \, a, b \in D\right\} \quad \text{and} \quad S_2 = \left\{\begin{pmatrix} 0 & a \\ 0 & b \end{pmatrix} \middle| \, a, b \in D\right\}.$$

Then S_1 and S_2 are left ideals, and the function $f: S_1 \to S_2$ defined by

$$f\begin{pmatrix} a & 0 \\ b & 0 \end{pmatrix} = \begin{pmatrix} 0 & a \\ 0 & b \end{pmatrix}$$

is an R-isomorphism since it is an isomorphism of the additive Abelian-group structure and for any $\begin{pmatrix} w & x \\ y & z \end{pmatrix}$ in R we have

$$f\left(\begin{pmatrix} w & x \\ y & z \end{pmatrix}\begin{pmatrix} a & 0 \\ b & 0 \end{pmatrix}\right) = f\begin{pmatrix} wa + xb & 0 \\ ya + zb & 0 \end{pmatrix} = \begin{pmatrix} 0 & wa + xb \\ 0 & ya + zb \end{pmatrix}$$

$$= \begin{pmatrix} w & x \\ y & z \end{pmatrix}\begin{pmatrix} 0 & a \\ 0 & b \end{pmatrix} = \begin{pmatrix} w & x \\ y & z \end{pmatrix} f\begin{pmatrix} a & 0 \\ b & 0 \end{pmatrix}.$$

We have the R-module decomposition $R = S_1 \oplus S_2$, and the S_i are simple: If either a or b is not zero then

$$R\begin{pmatrix} a & 0 \\ b & 0 \end{pmatrix} = S_1$$

since for any c and d in D we can find w, x, y, z in D so that

$$\begin{pmatrix} w & x \\ y & z \end{pmatrix}\begin{pmatrix} a & 0 \\ b & 0 \end{pmatrix} = \begin{pmatrix} c & 0 \\ d & 0 \end{pmatrix}.$$

We have seen that R is semisimple. Our discussion has been carried out for $D^{2 \times 2}$, but similar steps can be carried out for $D^{n \times n}$, where n is any positive integer. Thus, *for any division ring D and any positive integer n the ring $D^{n \times n}$ of all $n \times n$ matrices with entries in D is semisimple.* In more detail, $D^{n \times n}$ is the direct sum of n mutually isomorphic simple left ideals.

Let's return to $R = D^{2 \times 2}$. Since R is semisimple, we know that every R-module is the direct sum of simple modules and that each simple R-module is isomorphic with S_1. The same is true regarding $D^{n \times n}$. *For any division ring D and any positive integer n the ring $D^{n \times n}$ has, up to isomorphism, only one simple module S, and every module over $D^{n \times n}$ is a direct sum of copies of S.*

Now let D and E be division rings, and let $R = D^{2 \times 2} \times E^{3 \times 3}$, where the addition and multiplication in R are given by $(x, y) + (x', y') = (x + x', y + y')$ and $(x, y)(x', y') = (xx', yy')$ for all x, x' in $D^{2 \times 2}$ and all y, y' in $E^{3 \times 3}$. We call the ring R the *ring direct product* of $D^{2 \times 2}$ and $E^{3 \times 3}$. For the module R we can easily verify that $R = D^{2 \times 2} \oplus E^{3 \times 3} = S_{11} \oplus S_{12} \oplus S_{21} \oplus S_{22} \oplus S_{23}$, where

$$S_{11} = \left\{ \begin{pmatrix} a & 0 \\ b & 0 \end{pmatrix} \middle| \, a, b \in D \right\} \cong S_{12}$$

and

$$S_{21} = \left\{ \begin{pmatrix} a & 0 & 0 \\ b & 0 & 0 \\ c & 0 & 0 \end{pmatrix} \middle| \, a, b, c, \in E \right\} \cong S_{22} \cong S_{23} \not\cong S_{11}.$$

Thus, R is semisimple and each simple R-module is isomorphic with either S_{11} or S_{21}. Then every R-module is isomorphic to a direct sum of copies of S_{11} and (or) S_{21}. This argument holds for any finite direct product of total matrix rings over division rings. *For $R = D_1^{n(1) \times n(1)} \times D_2^{n(2) \times n(2)} \times \cdots \times D_k^{n(k) \times n(k)}$* (where $k, n(1), \ldots, n(k)$ *are positive integers and* $D_1, \ldots, D_k$ *are division rings*), *R is semisimple: R possesses exactly k mutually nonisomorphic simple modules, and every R-module is a direct sum of copies of certain of these simple modules.*

All of our assertions about direct products of total matrix rings have admitted very simple and straightforward verifications. The result that we wish to discuss now is considerably more tedious to demonstrate but is remarkable. It is essentially the converse of our previous discussion. *Every semisimple ring is isomorphic to the direct product of a finite number of total matrix rings over division rings.* We do not wish to prove this theorem here, but we will show how one can obtain from a semisimple ring the various division rings and the associated matrix dimensions required to build an isomorphic copy of the ring from total matrix rings. Let R be a semisimple ring, and suppose in particular that $R = S_{11} \oplus S_{12} \oplus S_{21} \oplus S_{22} \oplus S_{23} \oplus S_{31}$, where $S_{11} \cong S_{12} \not\cong S_{21} \cong S_{22} \cong S_{23} \not\cong S_{31} \not\cong S_{11}$ are simple left ideals. Let D_1, D_2, and D_3 be the opposite rings (Appendix 2 of Chapter 0) of the endomorphism rings (Exercise 12, Chapter 2) of S_{11}, S_{21}, and S_{31}, respectively. It is quite easy to verify, using the simplicity of the modules, that these rings are division rings (Exercise 12c, Chapter 2). A tedious investigation will show that $R \cong D_1^{2\times 2} \times D_2^{3\times 3} \times D_3^{1\times 1}$ (and of course $D_3^{1\times 1} \cong D_3$). This shows the origin of the division rings and the associated matrix dimensions. Our discussion has been given to indicate that the structure of semisimple rings may be regarded as thoroughly understood (if one assumes knowledge of the structure of division rings).

Reviewing Chapters 2 and 3 up to this point, we might summarize thus: if we are dealing with a ring for which all modules are free, then we are working with vector spaces, and we are in the thoroughly classical setting of linear algebra. If we are dealing with a ring for which all modules are projective, then we are again in a linear algebra setting; that is, we are concerned with the actions of total matrix rings on their simple left ideals. The problem of representing groups by means of nonsingular matrices over a ring is often regarded as a higher stage of linear algebra. Section 4 may therefore be regarded as a continuation of our incorporation of linear algebra into the general theory of modules.

§ 4 GROUP RINGS

There is an extremely important class of rings that are semisimple but for which their semisimplicity is not apparent. They are a particular class of what are called group rings. We will begin by defining the group ring of an arbitrary group G over an arbitrary ring R.

Ignore (for a moment) the algebraic structure on G, and regard G only as a set. Use G to index a family $\{R_\mu | \mu \in G\}$ of copies of R, and form the free R-module $R(G) = \oplus\{R\mu \mid \mu \in G\}$. Recall that the elements of this free module are the finitely nonzero functions $f : G \to R$. We will define a ring structure on $R(G)$. For the additive group we use the R-module addition of $R(G)$. We define the product of elements f and g in $R(G)$ to be the function $f \cdot g$ for which $f \cdot g(\mu) = \Sigma\{f(\alpha)g(\beta) \mid \alpha\beta = \mu\}$. Since both f and g are finitely nonzero, only finitely many of the products $f(\alpha)g(\beta)$ are nonzero; consequently, each sum $\Sigma\{f(\alpha)g(\beta) \mid \alpha\beta = \mu\}$ is defined, and $f \cdot g$ is also finitely nonzero. With a little patience the ring axioms can be verified for $R(G)$. *This ring $R(G)$ is called the group ring of the group G over the ring R.*

There is a natural and useful way to embed both G and R in $R(G)$. We have the usual basis $B = \{b_\mu \mid \mu \in G\}$ of the free R-module $R(G)$, where each b_μ is defined by $b_\mu(v) = 1$ if $v = \mu$ and $b_\mu(v) = 0$ if $v \neq \mu$. It is easy to see that B is closed under the multiplication of $R(G)$ and that the function $h : G \to R(G)$ defined by $h(\mu) = b_\mu$ is an isomorphism of G with B. The identity element of the ring $R(G)$ is the basis element b_1, where 1 denotes the identity element of G. The function $k : R \to R(G)$ defined by $k(r) = rb_1$ is an isomorphism of the ring R with the subring $\{rb_1 \mid r \in R\}$ of $R(G)$. We will use the homomorphisms h and k to embed G and R in $R(G)$. By this we mean that we shall write for each b_μ simply μ and thereby consider that G is a subset of $R(G)$. For each rb_1 we shall write simply r and thereby consider that R is a subset of $R(G)$. Let us see what this change of notation does for an arbitrary f in $R(G)$. There is a unique indexed family $\{r_\mu \mid \mu \in G\}$ of elements of R for which $f = \Sigma r_\mu b_\mu$. This expression may now be rewritten $f = \Sigma r_\mu b_\mu = \Sigma r_\mu(b_1 b_\mu) = \Sigma(r_\mu b_1)b_\mu = \Sigma r_\mu \mu$, where each r_μ and each μ are regarded as elements of $R(G)$ and where each term $r_\mu \cdot \mu$ is regarded as a product formed with respect to the multiplication of $R(G)$.

Since R is a subring of $R(G)$, every $R(G)$-module may also be regarded as an R-module. If $f : A \to B$ is an R-homomorphism between $R(G)$-modules and $f(\mu a) = \mu f(a)$ holds for all μ in G and all a in A, then f is an $R(G)$-homomorphism by the following calculation: $f((\Sigma r_\mu \mu)a) = f(\Sigma(r_\mu \mu a)) = \Sigma(r_\mu f(\mu a)) = \Sigma(r_\mu \mu f(a)) = (\Sigma r_\mu \mu) f(a)$. This fact and one more will be useful in the proof of Observation 3. Each element r in R commutes with each element μ in G: $r \cdot \mu = (rb_1)b_\mu = rb_\mu = (rb_\mu)b_1 = b_\mu(rb_1) = \mu \cdot r$.

OBSERVATION 3. If G is a finite group of order n and R is a division ring of characteristic either zero or a prime not dividing n, then $R(G)$ is a semisimple ring, and consequently every $R(G)$-module is semisimple.

Verification: To show that a ring is semisimple it is sufficient to show that each submodule of each module is a summand (Exercise 2).

Thus, we proceed by letting A be an arbitrary submodule of an arbitrary $R(G)$-module B. Let $i : A \to B$ be the inclusion map ($i(a) = a$). We need only find a map $h : B \to A$ for which hi is the identity map on A, since we will then have $B = A \oplus \ker(h)$. Since $R \subseteq R(G)$, B is also an R-module and A is an R-submodule. Since R is a division ring, there is an R-homomorphism $f : B \to A$ for which fi is the identity map on A. There is an easy way to modify f to produce an $R(G)$-homomorphism: First observe that for every μ in G, $\mu^{-1}f(\mu b)$ is an element of A. Thus, for each μ in G, we have a function $\mu^{-1}f\mu : B \to A$ defined for each b in B by $\mu^{-1}f\mu(b) = \mu^{-1}f(\mu b)$. Each $\mu^{-1}f\mu$ is an R-homomorphism. Summing these functions gives another R-homomorphism $k = \Sigma\{\mu^{-1}f\mu \mid \mu \in G\}$ of B into A. This k is actually an $R(G)$-homomorphism: For each v in G and each b in B we have $k(vb) = \Sigma\{\mu^{-1}f(\mu(vb)) \mid \mu \in G\} = v\Sigma\{(\mu v)^{-1}f((\mu v)b) \mid \mu \in G\} = v\Sigma\{\lambda^{-1}f(\lambda b) \mid \lambda \in G\} = vk(b)$. The second from the last equality follows from the fact that, for a fixed element v in a group, as μ runs through all the elements of the group without repetition, μv does likewise. For a in A we have $ki(a) = k(a) = na$, where n is the order of G. Since the characteristic of R is either zero or a prime not dividing n, n has an inverse n^{-1} in R. We may now define $h = n^{-1}k$ to obtain the desired $R(G)$-homomorphism $h : B \to A$ for which hi is the identity map on A.

The correspondence between matrix representations of groups and modules over group rings is outlined in a sequence of exercises below.

NOTES FOR CHAPTER 3

The structural description of semisimple rings in terms of matrices over division rings that is discussed in § 3 is one of the major classical results of Wedderburn. We do not need a proof of this theorem for our work, but, to the student who would like to have one, I suggest the beautiful lecture by Emil Artin, "The Influence of J. H. M. Wedderburn on the Development of Modern Algebra" [1950]. Other proofs are available in Burrow[1965] and Lambek [1966].

In § 4 and Exercises 14 through 17 I have included the minimal amount of information needed to indicate how group-representation theory can be subsumed under the general structure theory of modules. The representations of a group G in terms of matrices over a field F need only be studied "up to equivalence." Thus, according to Exercises 14 through 17, the study of representations of G coincides with the study of the structure "up to isomorphism" of $F(G)$-modules. In the most important classical situations, G is a finite group and F is a subfield of the complex numbers. In such situations Observation 3 applies and tells us that in the final analysis we need only study the nonisomorphic simple left ideals of the semisimple ring $F(G)$. These will be finite in number and will yield all of the so-called "inequivalent irreducible representations" from which all representations can be assembled. As you see, what we have listed as "Observation 3" is an extremely powerful result. It is a classical theorem of algebra due to H. Maschke [1898].

If you would like to see group-representation theory discussed in the context of module theory, I suggest Curtis and Reiner [1962] or Burrow [1965]

EXERCISES

1. a. For which nonnegative integers n is the ring Z_n of integers modulo n a semisimple ring?
   b. Describe the structure of all Z_{30}-modules.
   c. Describe a Z_{150}-module that is not projective.
   d. If a ring R has a nonprojective module, must it have a nonprojective cyclic module? Must it have a nonprojective left ideal?
2. Prove that a ring R is semisimple if and only if each submodule S of each R-module A is a direct summand of A.
3. a. Prove that a ring R is semisimple if and only if each (maximal) left ideal of R is a direct summand of R.
   b. Must a ring be semisimple if each of its left ideals is a principal module?
4. Verify that, for each r in the ring Q of rational numbers, the subset

$$S_r = \left\{ \begin{pmatrix} a & ar \\ b & br \end{pmatrix} \middle| \ a, b \in Q \right\}$$

of $Q^{2\times 2}$ is a simple left ideal.

5. For the ring $Q^{3\times 3}$ we have $Q^{3\times 3} = S_1 \oplus S_2 \oplus S_3$, where each S_i consists of those matrices whose entries outside the ith column are zero. Let

$$L = \left\{ \begin{pmatrix} 0 & a & a \\ 0 & b & b \\ 0 & c & c \end{pmatrix} \middle| \ a, b, c \in Q \right\}.$$

   a. Verify that L is a left ideal of $Q^{3\times 3}$.
   b. Carry out the procedure described in the verification of Observation 2 and list the elements of the resulting set F.

c. Give a simple description of the left ideal of $Q^{3\times 3}$ that is generated by

$$\begin{pmatrix} 1 & 0 & 2 \\ 1 & 0 & 2 \\ 1 & 0 & 1 \end{pmatrix}.$$

6. From the discussion given in § 3, what can be said about the structure of commutative semisimple rings?

7. Let M be a semisimple R-module.
   a. Is every maximal simply independent subset of M a maximal independent subset of M?
   b. Is every independent subset simply independent?
   c. Does every maximal (simply) independent subset of M necessarily generate M?

8. Prove the following assertions concerning the class of semisimple modules over an arbitrary ring R:
   a. Each submodule of each semisimple module is a direct summand.
   b. Each submodule of each semisimple module is semisimple.
   c. Each quotient module of each semisimple module is semisimple.

9. Alter the wording of the proof of Observation 2 of Chapter 2 to produce a proof of the following assertion: if M is a semisimple R-module having a maximal simply independent set B of infinite cardinal β, then each subset A of M having cardinal $\alpha < \beta$ is contained in a proper direct summand of M.

10. Show that, if a semisimple R-module has an infinite (maximal simply) independent subset, then every pair of maximal simply independent subsets have the same cardinal.

11. Alter the wording of the proof of Observation 6 of Chapter 2 to produce a proof of the following assertion: two maximal simply independent subsets of the same semisimple module A over a ring R have the same cardinal number.

12. Let $\{S_i \mid i \in I\}$ be a set of simple R-modules for which $S_i \cong S_j$ only if $i = j$. Let M be an R-module for which $\oplus\{A_i \mid i \in I\} = M = \oplus\{B_i \mid i \in I\}$ where, for each i in I, A_i and B_i are direct sums of (possibly empty) families of submodules each isomorphic with S_i. Prove that $A_i = B_i$ for each i in I.

13. Let $\{S_i \mid i \in I\}$ be a family of simple R-modules with the property that each simple R-module is isomorphic with precisely one S_i. For each i in I, let $\{S_{ij} \mid j \in J(i)\}$ and $\{S'_{ik} \mid k \in K(i)\}$ be (possibly empty) families of R-modules for which $S_{ij} \cong S_i \cong S'_{ik}$ hold for all j in $J(i)$ and k in $K(i)$. Using the results of the two previous exercises, prove that $\oplus\{S_{ij} \mid i \in I, j \in J(i)\}$ and $\oplus\{S'_{ik} \mid i \in I, k \in K(i)\}$ are isomorphic (if and) only if, for each i in I, $J(i)$ and $K(i)$ have the same cardinal number.

14. Let F be a field, and let $V = F^n$ be the usual vector space (that is, F-module) of ordered n-tuples of elements of F. Let $M = F^{n\times n}$ be the ring of $n \times n$ matrices

with entries from F. Let G be a group and let $h : G \to M$ be a function such that $h(\sigma)$ is nonsingular for each σ in G and for which $h(\sigma\tau) = h(\sigma)h(\tau)$ for all σ, τ in G. Homomorphisms such as h are called *group representations.* Define an operation of the group ring $F(G)$ on V by means of $(\Sigma\{r_\mu \mu \mid \mu \in G\}) \cdot v = \Sigma\{r_\mu \cdot h(\mu) \cdot v \mid \mu \in G\}$. Verify that with this operation V becomes an $F(G)$-module.

15. Let V be an $F(G)$-module, where F is a field and G is a group. Since we regard F as a subset of $F(G)$, V also possesses a vector space structure over F. Assume that V has a finite dimension n as a vector space, and let $b_1, \ldots, b_n$ be a basis of V. Each σ in G provides a function $m_\sigma : V \to V$ defined by $m_\sigma(v) = \sigma \cdot v$.
    a. Verify that m_σ is an F-homomorphism. Let $M = F^{n \times n}$, and let $k : G \to M$ be the function defined by specifying that, for each σ in G, $k(\sigma)$ is the matrix of the linear mapping m_σ formed with respect to the (ordered) basis $b_1, \ldots, b_n$.
    b. Verify that for each σ in G, $k(\sigma)$ is nonsingular.
    c. Verify that $k(\sigma \cdot \tau) = k(\sigma) \cdot k(\tau)$ holds for all σ, τ in G and therefore that k is a group representation in the sense of Exercise 14.

16. a. Let F, V, M, G, and h be as in Exercise 14, and regard V as an $F(G)$-module as described in that exercise. Using the usual basis choice $b_1 = (1, 0, \ldots, 0), \ldots, b_n = (0, \ldots, 0, 1)$ in V, a group respresentation $k : G \to M$ is provided by Exercise 15. Verify that $k = h$.
    b. Let V, F, G, and M be as in Exercise 15, and let k be the group representation described in that exercise. Placing k in the role of the function h in Exercise 14, an $F(G)$-module structure is provided for V by that exercise. Verify that this $F(G)$-module structure is identical with the originally given $F(G)$-module structure of V.

17. Let F, V, M, and G be as in Exercise 14 and let both $h_1 : G \to M$ and $h_2 : G \to M$ be group representations in the sense of that exercise. Then h_1 and h_2 are said to be *equivalent representations* if there exists a nonsingular matrix T in M for which $h_1(\sigma) = Th_2(\sigma)T^{-1}$ for all σ in G. In accord with Exercise 14, h_1 provides an $F(G)$-module structure for V and so does h_2. Prove that h_1 and h_2 are equivalent if and only if they provide isomorphic $F(G)$-module structures for V.

18. Prove that a uniform projective R-module must be isomorphic with a left ideal of R.

CHAPTER 4

SUBMODULES OF FREE MODULES

§ 1 INTRODUCTION

We have made a preliminary investigation of summands of free modules. What can we say about submodules of free modules? We have one elementary tool that can be used to analyze such submodules.

OBSERVATION 1. Let S be a submodule of the module $A \oplus B$, and suppose that $p(S)$ is projective, where p is the natural projection of $A \oplus B$ onto B. Then $S = (S \cap A) \oplus B'$ for a submodule $B' \cong p(S)$.

Verification: Let p' be the restriction of p to S. Since $\operatorname{im}(p') = p(S)$ is projective, we know by Observation 4 of Chapter 2 that p' splits and therefore that $S = \ker(p') \oplus B'$ for a submodule $B' \cong \operatorname{im}(p')$. Since $\ker(p') = S \cap A$, the verification is complete.

The manner in which projectivity is involved suggests that the observation can only be used to analyze submodules of free modules over rings with the property that all left ideals are projective. These rings will be introduced in the next section.

§ 2 HEREDITARY RINGS

The broadest class of rings for which we will be able to describe all submodules of free modules is the class described as follows.

DEFINITION. A ring R is *hereditary* if each of its left ideals is projective.

All semisimple rings are hereditary. The two classes of rings can be compared most readily through an alternate definition of semisimple rings: a ring R is semisimple if each of its left ideals is a summand. The ring Z of integers is hereditary since each nonzero ideal is of the form $nZ \cong Z$. Similarly, the ring $D[x]$ of polynomials in x with coefficients in the division ring D can be shown to be hereditary.

We will approach the description of the submodules of free modules over hereditary rings very gradually by means of two observations.

OBSERVATION 2. Let R be hereditary, and let S be a submodule of the free R-module $R_0 \oplus \cdots \oplus R_{n-1}$. Then $S = A_0 \oplus \cdots \oplus A_{n-1}$, where each A_i is isomorphic with a left ideal of R.

This observation can be proved by finite induction on n. For $n = 1$ the assertion holds for arbitrary rings, and the required induction step is provided by Observation 1. We will not give the details since they are present in the verification of Observation 3, which may be interpreted to subsume Observation 2.

OBSERVATION 3. Let R be hereditary, and let S be a submodule of the free R-module $F = \oplus\{R_i \mid i$ a nonnegative integer$\}$. Then $S = \oplus\{A_i \mid i$ a nonnegative integer$\}$, where each A_i is isomorphic with a left ideal of R.

Verification: For each nonnegative integer j, let $S_j = S \cap \oplus\{R_i \mid 0 \leq i \leq j - 1)$. Then $S_0 = 0$ and by the nature of a direct sum (as opposed to a direct product), $S = \cup\{S_i \mid i$ a nonnegative integer$\}$. For each j let p_j be the projection of $\oplus R_i$ onto R_j and let p'_j be the restriction of p_j to

S_{j+1}. For each j, $p'_j(S_{j+1})$ is a left ideal and is therefore projective. Since the kernel of p'_j is S_j, $S_{j+1} = S_j \oplus A_j$ for some submodule $A_j \cong p'_j(S_{j+1})$. We note that $A_0 = S_1$ and observe that each A_i is isomorphic with a left ideal of R. We will show that $S = \oplus\{A_i \mid i \text{ a nonnegative integer}\}$. If this sum were not direct, a contradiction could be produced as follows: if $0 = a_{i(1)} + \cdots + a_{i(n)}$ with $i(1) < \cdots < i(n)$, $a_{i(j)} \in A_{i(j)}$, and $a_{i(n)} \neq 0$, then the directness of $S_{i(n)+1} = S_{i(n)} \oplus A_{i(n)}$ would be contradicted. If $S \neq \oplus A_i$, then there would exist a least k for which S_k is not contained in $\oplus A_i$. Since $S_0 = 0$, k could not be zero and the following contradiction of the choice of k would arise: $\oplus A_i \supseteq S_{k-1} \oplus A_{k-1} = S_k$.

Observations 2 and 3 show that a submodule of a free module F over a hereditary ring must be isomorphic with a direct sum of left ideals if F has a finite or countably infinite basis. In § 3 we will remove this cardinality restriction on F.

§ 3 (Ord.) HEREDITARY RINGS (CONTINUED)

The key to the analysis of the submodules of free modules over hereditary rings is Observation 1. To make the analysis, we need only iterate the procedure given by this observation. Observations 2 and 3 were made by a finite iteration and a countably infinite iteration, respectively. What we need now is a vehicle that will allow iteration beyond the countable realm. For this purpose we will use ordinal numbers. For an understanding of the ordinal numbers and their elementary properties, I suggest the book by P. R. Halmos [1960]. The proof of Theorem 2 has been written to invite detailed comparison with the proof of Observation 3. The comparison will pinpoint the few specifically new features involved in iterating beyond the countable realm. In the initial step of the proof of the theorem we will use the following equivalent of the well-ordering principle: for every set I there is an ordinal number μ and a bijection $f: I \to \{\alpha \mid \alpha \text{ is an ordinal} < \mu\}$.

THEOREM 2. Each submodule of each free module over a hereditary ring is isomorphic to a direct sum of left ideals.

Proof: Let R be a hereditary ring. Each free R-module is isomorphic to a module $\oplus\{R_\alpha \mid \alpha \text{ is an ordinal number } < \mu\}$ for some ordinal number μ. (The free modules in Observations 2 and 3 have this form for $\mu = n$ and $\mu = \omega$, respectively.) Let $\oplus\{R_\alpha \mid \alpha < \mu\}$ be such an R-module, and let S be a submodule. For each ordinal $\beta \leq \mu$, let $S_\beta = S \cap \oplus\{R_\alpha \mid \alpha < \beta\}$. Then $S_0 = 0$ and $S_\mu = S$. For each $\beta < \mu$, let p_β be the projection of $\oplus\{R_\alpha \mid \alpha < \mu\}$ onto R_β and let p'_β be the restriction of p_β to $S_{\beta+1}$. For each $\beta < \mu$, $p'_\beta(S_{\beta+1})$ is a left ideal and is therefore projective. Since the kernel of p'_β is S_β, $S_{\beta+1} = S_\beta \oplus A_\beta$ for some submodule $A_\beta \cong p'_\beta(S_{\beta+1})$. We note that $A_0 = S_1$ and observe that each A_α is isomorphic with a left ideal of R. We will show that $S = \oplus\{A_\alpha \mid \alpha < \mu\}$. If the sum were not direct, a contradiction could be produced as follows: if $0 = a_{\alpha(1)} + \cdots + a_{\alpha(n)}$ with $\alpha(1) < \cdots < \alpha(n) < \mu$, $a_{\alpha(i)} \in A_{\alpha(i)}$, and $a_{\alpha(n)} \neq 0$, then the directness of $S_{\alpha(n)+1} = S_{\alpha(n)} \oplus A_{\alpha(n)}$ would be contradicted. If $S \neq \oplus A_\alpha$, then there would exist a least ordinal $\lambda \leq \mu$ for which S_λ is not contained in $\oplus A_\alpha$. Since $S_0 = 0$, λ could not be zero. If λ has a predecessor, we have $S_{\lambda-1} \subseteq \oplus A_\alpha$ and $S_\lambda = S_{\lambda-1} \oplus A_{\lambda-1} \subseteq \oplus A_\alpha$. If λ has no predecessor, then we have $S_\lambda = \cup\{S_\beta \mid \beta < \lambda\}$, and the inclusions $S_\beta \subseteq \oplus A_\alpha$, which hold for all $\beta < \lambda$, give $S_\lambda \subseteq \oplus A_\alpha$. Thus, a contradiction of the choice of λ arises whether λ has a predecessor or not. We conclude that $S = \oplus\{A_\alpha \mid \alpha < \mu\}$ as required.

This theorem has four corollaries. The first is a structure theorem for projective modules over hereditary rings.

COROLLARY 1. A projective module over a hereditary ring is isomorphic to a direct sum of left ideals.

The second and third corollaries are propositions that characterize hereditary rings.

COROLLARY 2. A ring is hereditary if and only if each submodule of each of its free modules is projective.

The third corollary explains the choice of the term "hereditary."

COROLLARY 3. A ring is hereditary if and only if each submodule of each of its projective modules "inherits" the projectivity of the containing module.

The final corollary provides an introduction to the next section.

COROLLARY 4. A ring has the property that each submodule of each of its free modules is free if and only if each of its left ideals is free.

§ 4 PRINCIPAL IDEAL DOMAINS

If we ask which *commutative rings* R have the property that submodules of free R-modules are free, we arrive at one of the most widely used classes of rings.

DEFINITION. A ring R is a *principal ideal domain* if R is a commutative integral domain and each ideal of R is cyclic.

The most common examples of principal ideal domains are the ring of integers and the ring $F[x]$ of polynomials in one variable x with coefficients in a field F. Each field is also a principal ideal domain.

OBSERVATION 4. The nonzero free ideals of a commutative ring R are precisely those ideals of the form Rr, where r is an element of R which is not a zero divisor.

Verification: Let F be a nonzero free ideal of a commutative ring R. Let r be an element of any basis of F. Then r is not a zero divisor since $r'r = 0$ and $r' \neq 0$ would violate the defining property of the basis to which r belongs. If F has a basis consisting of two or more elements, say r and r', then we have $F = Rr \oplus Rr' \oplus F'$, where F' is the submodule generated by the remaining elements of the basis to which r and r' belong. But then $r'r = rr' \in Rr \cap Rr' = 0$, which yields $r'r = 0$, contradicting the fact that r cannot be a zero divisor. Thus, $F = Rr$, where r is not a zero divisor.

Let r be an element of R which is not a zero divisor. Then the kernel of the map $h : R \to Rr$ defined by $h(r') = r'r$ must be zero and $Rr \cong R$ is free.

PROPOSITION 3. A commutative ring R has the property that each submodule of each free R-module is free if and only if R is a principal ideal domain.

Proof: Suppose R is a principal ideal domain. Then each nonzero ideal is of the form Rr, where r cannot be a zero divisor. Thus, every ideal of R is free, and by Corollary 4 of § 3 submodules of free R-modules are necessarily free.

Suppose that R is a commutative ring for which submodules of free R-modules are necessarily free. Then every ideal of R is free; consequently, by our observation above, every nonzero ideal of R is of the form Rr for some element r in R that is not a zero divisor. We need only show that R is an integral domain. Suppose u and v are nonzero elements of R for which $uv = 0$. But $Rv = Rr$ for some r in R that is not a zero divisor. Then there is a w in R such that $r = wv$, and we have $ur = uwv = 0$, contradicting the assumption that r is not a zero divisor. We conclude that R is a principal ideal domain.

For which rings R is every R-module isomorphic with a submodule of a free module? The answer to this question is the class of quasi-Frobenius rings, which will be discussed briefly in § 5 of Chapter 9. A proof will not be given in this book, but the problem will be discussed in the notes following Chapter 9.

NOTES FOR CHAPTER 4

One of the central topics of linear algebra is the determination of the similarity classes of linear transformations. If we restrict our attention to linear transformations on finite dimensional vector spaces and express the problem in matrix form, then we have the usual problem of finding the similarity classes of $n \times n$ matrices over a field. What is desired is then a collection of $n \times n$ matrices over the field that has the property that each $n \times n$ matrix is similar to exactly one matrix of the collection. The similarity theory of matrices leads to the rational canonical form and Jordon canonical form for matrices. In Exercises 6 through 10 I have included the minimal amount of information needed to indicate how the theory of similarity of linear transformations (and matrices) can be subsumed under the general structure theory of modules. If you would like to see this portion of linear algebra discussed in the context of module theory, I suggest Hartley and Hawkes [1970], MacLane and Birkhoff [1967], or Rotman [1965].

EXERCISES

1. Let J be a nonzero ideal of the ring Z of integers, and let g be a nonzero element of J having least absolute value.
   a. Verify that $J = Zg$.

b. Verify that the function $f : Z \to Zg$ defined by $f(z) = z \cdot g$ is an isomorphism of Z-modules and consequently that Z is hereditary.

2. Let D be a division ring, and let $D[x]$ be the ring of polynomials in x with coefficients in D. Let J be a nonzero ideal of $D[x]$, and let $g(x)$ be a nonzero element of J having least degree.
   a. Verify that $J = D[x] \cdot g(x)$.
   b. Verify that the function $f : D[x] \to D[x] \cdot g(x)$ defined by $f(p(x)) = p(x) \cdot g(x)$ is an isomorphism of $D[x]$-modules and consequently that $D[x]$ is hereditary.

3. Prove Observation 2, using Observation 1 and finite induction.

4. a. Study the proof of Observation 3 to determine whether the submodules A_i of S are actually subsets of the corresponding submodules R_i of F.
   b. Let Z be the ring of integers, and let F be the free Z-module $F = \oplus\{Z_i \mid i \text{ a nonnegative integer}\}$. Let S be the submodule of F consisting of those functions (sequences) for which $f(0) = f(1)$ and $f(i) = 0$ for $i \geq 2$. The proof of Observation 3 provides submodules A_i of S for which $S = \oplus\{A_i \mid i \text{ a nonnegative integer}\}$. For each i, state specifically which elements of S lie in A_i.

5. Let R be an arbitrary ring. An R-module H is said to be a *hereditary* R-module if M and each of its submodules is projective.
   a. Modify the proof of Observation 3 to obtain the following generalization: let $\{H_i \mid i \text{ a nonnegative integer}\}$ be a family of hereditary R-modules, and let S be a submodule of $\oplus H_i$. Then $S = \oplus A_i$, where each A_i is isomorphic with a submodule of the corresponding H_i.
   b. Show by example that, if the words "isomorphic with" are deleted from the previous sentence, then the resulting sentence is false.
   c. Modify the statement and proof of Theorem 2 to obtain a generalization concerning submodules of direct sums of hereditary modules.
   d. Show that the direct sum of any family of hereditary R-modules is hereditary (that is, that the semihereditary R-modules are precisely the hereditary R-modules).

6. Let A be a vector space over the field F, and let $t : A \to A$ be a linear transformation (that is, an F-homomorphism of A). Define an operation of the ring $F[x]$ on A by means of $(s_0 + s_1x + \cdots + s_nx^n) \cdot a = s_0 \cdot a + s_1 \cdot t(a) + \cdots + s_n \cdot t^n(a)$. Verify that this operation of $F[x]$ on A provides an $F[x]$-module structure for A.

7. Let F be a field, and let A be an $F[x]$-module. Since F is a subring of $F[x]$, A is also an F-module (that is, a vector space over the field F). Let $t : A \to A$ be the function defined by $t(a) = x \cdot a$. Verify that t is an F-homomorphism (that is, a linear transformation of the vector space A).

8. a. Let A, F, and t be as in Exercise 6, and regard A as an $F[x]$-module as described in that exercise. From this $F[x]$-module structure, a linear transformation $t' : A \to A$ is obtained as described in Exercise 7. Verify that $t' = t$.

b. Let F, A, and t be as in Exercise 7. Regard A as a vector space over F, and use t to provide A with an $F[x]$-module structure as in Exercise 6. Verify that this $F[x]$-module structure is identical with the originally given $F[x]$-module structure of A.

9. Let A be a vector space over the field F, and let $t_1 : A \to A$ and $t_2 : A \to A$ be linear transformations of A. Then t_1 and t_2 are said to be *similar* if there exists a nonsingular linear transformation $u : A \to A$ for which $t_1 = ut_2u^{-1}$. In accord with Exercise 6, t_1 provides an $F[x]$-module structure for A and so does t_2. Prove that t_1 and t_2 are similar if and only if they provide isomorphic $F[x]$-module structures for A.

10. a. Explain how Exercises 6 through 9 give precise meaning to the following assertion and how they justify this assertion: the problem of classifying linear transformations "up to similarity" is a special case of the problem of classifying modules over principal ideal domains "up to isomorphism."

    b. Let $F^{n \times n}$ be the ring of all $n \times n$ matrices over the field F. Matrices M_1 and M_2 in $F^{n \times n}$ are said to be *similar* if there is a nonsingular matrix U in $F^{n \times n}$ for which $M_1 = UM_2U^{-1}$. Explain how Exercises 6 through 9 justify the following assertion: the problem of classifying the matrices in $F^{n \times n}$ "up to similarity" is equivalent to the problem of classifying "up to isomorphism" the $F[x]$-modules that are n-dimensional when regarded as a vector space over the subfield F of $F[x]$.

CHAPTER 5

QUOTIENT MODULES OF FREE MODULES AND GENERATING SETS

Through Part One we have expanded our attention successively from free modules to summands of free modules to submodules of free modules. In the present chapter we discuss the quotient modules of free modules. This discussion represents the ultimate expansion of our attention, since, as we shall see, *every module is isomorphic to a quotient module of a free module.* This fact was implicit in the proof of Observation 1 of Chapter 3. We will give a more thorough discussion here.

Let R be a ring and A an R-module. We will survey the ways of representing A as a homomorphic image of a free module. The process can be organized into three choices: (1) Choose a set G of generators of A. (2) Choose a free module F having a basis B of cardinal greater than or equal to the cardinal of G. (3) Choose a surjective function $f: B \to G$, and let $h: F \to A$ be the map determined by f. Notice that Choice (2) is always possible, since for every cardinal number there is a free module with a basis of that cardinal (see Chapter 2, § 2). The map h is necessarily an epimorphism since its image contains the set G, which generates A.

Notice that every epimorphism $h' : F' \to A$ of a free module F' onto A may be regarded as having arisen from such choices: let B be a basis of F', G be $h'(B)$, and f be the restriction of h' to B. Then h' is the map corresponding to the choice of G as the generating set of A, F' as the free module having B as the appropriate basis, and f as the surjective function $f : B \to G$.

Which free modules possess A as a homomorphic image? To give the answer we associate a cardinal number α with A. Let α be the least cardinal number in the set $\{\kappa \mid \kappa$ is the cardinal number of some set of generators of $A\}$. Such an α exists since every set of cardinals contains a least element. From the previous paragraph we may conclude that *A is a homomorphic image of the free module F if and only if F possesses a basis of cardinal $\beta \geq \alpha$.* In particular, each free module having a basis of cardinal α is, in a sense, a "smallest" free module having A as a homomorphic image.

The program we have laid down for Part One has come to a natural resting point. Our basic structural theme has been projectivity: all the rings that have arisen in our study are hereditary, and all the modules for which we have structural results are projective. In Part Three we pick up the theme of projectivity again with our concern tempered by a sensitivity to the cardinality of generating sets. In the notation of our discussion above, Part Three begins with an examination of modules A for which α is finite and proceeds to the study of those for which α is countable. The reader who has an urgent desire for more information concerning projectivity may proceed immediately to Chapters 11 through 13, which are independent of Chapters 6 through 10. The general reader should begin Part Two, which introduces a new theme (injectivity) that requires fresh patterns of thought.

NOTES FOR CHAPTER 5

I will review Part One in two paragraphs that emphasize as strongly as possible the connection of Part One with the material of introductory modern algebra courses.

The development of Part One has introduced (or reintroduced) the most familiar classes of rings. Division rings arose in Chapter 2 as the class of rings for which all modules are free. The rings of $n \times n$ matrices over division rings (and finite direct products of them) arose in Chapter 3 as the class of rings for which all modules are projective. The hereditary rings that arose in Chapter 4 are admittedly not discussed in introductory modern algebra courses, but along with these rings came the principal ideal domains, which arose as the class of commutative rings for which submodules of free modules are free. Thus, the "bread and butter" rings of elementary algebra arose naturally in the initial steps of the study of the structure of modules.

Consider Part One in relation to the subject matter of linear algebra. Vector spaces arose in Chapter 2. The rings of linear transformations of finite dimensional vector spaces arose in Chapter 3. The study of similarity of linear transformations of vector spaces over a field F is equivalent to the study of the structure of modules over the principal ideal domain $F[x]$. Initial steps have been made in the study of such modules in Chapter 4. Finally, the study of representations of a group G by linear transformations of vector spaces over a field F is equivalent to the study of modules over the group ring $F(G)$. Such rings and their modules were the subject of § 4 of Chapter 3.

I hope that Part One has given you the feeling that the problem of describing the structure of modules is interwoven with traditional concerns of modern algebra. In Parts Two and Three we will head into the study of the structure of modules, treating the subject as one having intrinsic interest. However, if your future interests take you on into the study of homological algebra, you will find that every topic we take up in Parts One, Two, and Three will be of value in your later studies.

EXERCISES

1. Prove that an R-module P is projective if and only if for each epimorphism $h: A \to P$ there is a splitting map (that is, a map $s: P \to A$ for which hs is the identity map on P).

2. Let Z_m and Z_n be the Z-modules of integers mod m and n, respectively. What is the value of α for $Z_m \oplus Z_n$?

3. Regard the additive group Q of rational numbers as a Z-module.
   a. What is the value of α for this module?
   b. Show that each set of generators for Q contains a proper subset that also generates Q.

4. Let A be an R-module for which α is infinite, and let G be any subset of A that generates A. Show that G contains a subset of cardinal number α that generates A.

5. a. Let A be an R-module for which α is finite, and let G be a subset of A that generates A. Show that G contains a finite subset S that generates A.
   b. Let n be a positive integer. Find a cyclic Z-module A that contains a generating set G consisting of n elements and containing no proper subset that generates A.

6. Find a ring having the property that each of its finitely generated modules is cyclic.

PART TWO

INJECTIVITY

CHAPTER 6

ENLARGING HOMOMORPHISMS

§ 1 INTRODUCTION

A basic theme in Part One was the lifting of maps through epimorphisms. The basic theme in Part Two is dual, in a diagrammatic sense, to the lifting problem. Let A be a submodule of an R-module B, and let $h : A \to C$ be a map of A into an R-module C. When can a map $k : B \to C$ be found that agrees with h on A?

We will begin by considering a more general question. For A an R-submodule of B and a map $h : A \to C$, in what circumstances can a submodule A' and a map $h' : A' \to C$ be found that satisfy $A \subset A' \subseteq B$ and $h' \mid A = h$? The simplest result is stated in Observation 1.

OBSERVATION 1. If A and X are submodules of an R-module B and $h : A \to C$ and $k : X \to C$ are maps which agree on $A \cap X$ (that is, $h \mid A \cap X = k \mid A \cap X$), then they have a common enlargement $h' : A + X \to C$ (that is, $h' \mid A = h$ and $h' \mid X = k$).

Verification: For $a + x \in A + X$ we attempt a definition: $h'(a + x) = h(a) + k(x)$. It is clear that h' will be a map with the desired properties if it is a well-defined function. If $a + x = a' + x'$, then $a - a' = x' - x \in A \cap X$ and $h(a - a') = k(x' - x)$. The latter equation gives $h(a) - h(a') = k(x') - k(x)$ and $h(a) + k(x) = h(a') + k(x')$, which confirms that h' is well defined.

We will expand our field of attention to the set of *all* maps h': $A' \to C$ for which $A \subseteq A' \subseteq B$ and $h' \mid A = h$. The graph $\mathrm{gr}(h')$ of each such h' is a submodule of $B \oplus C$ containing $\mathrm{gr}(h)$ for which $p_B \mid \mathrm{gr}(h')$ is a monomorphism (with image A'). Conversely, any submodule S of $B \oplus C$ that contains $\mathrm{gr}(h)$ and satisfies the condition that $p_B \mid S$ is a monomorphism is the graph of a map h' of $p_B(S)(\supseteq A)$ into C that satisfies $h' \mid A = h$. The family of submodules $\{\mathrm{gr}(h') \mid h' \text{ is an enlargement of } h\}$ of $B \oplus C$ must contain a maximal nest $\mathcal{M}$. Let M be the union of such a maximal nest. Now $M = \mathrm{gr}(h')$ for an enlargement h' of h, according to the following considerations. Since M is the union of a nest of submodules containing $\mathrm{gr}(h)$, M is a submodule containing $\mathrm{gr}(h)$. For any pair (b, c), (b', c) of elements of M, there is a k such that $\mathrm{gr}(k)$ is a member of the nest $\mathcal{M}$ of which M is the union, and $\mathrm{gr}(k)$ contains both (b, c) and (b', c). Since $p_B \mid \mathrm{gr}(k)$ is a monomorphism, $b = b'$. We conclude that $p_B \mid M$ is a monomorphism and $M = \mathrm{gr}(h')$, where the domain of h' is $A' = p_B(M)$; for each x in A', $h'(x)$ is that element y of C for which $(x, y) \in M$. Does there exist a map $h'' : A'' \to C$ for which $A' \subset A'' \subseteq B$ and $h'' \mid A' = h'$? There does not, because otherwise, by adjoining $\mathrm{gr}(h'')$ to the nest $\mathcal{M}$, we would produce a strictly larger nest in contradiction of the maximality of $\mathcal{M}$. We have proved Observation 2.

OBSERVATION 2. For any R-homomorphism $h: A \to C$, where A is a submodule of an R-module B, there is a submodule A' satisfying $A \subseteq A' \subseteq B$ and a map $h' : A' \to C$ satisfying $h' \mid A = h$; for this submodule A' there is no map $h'' : A'' \to C$ satisfying $A' \subset A'' \subseteq B$ and $h'' \mid A' = h'$.

We will refer to any such h' as *a maximal enlargement of h.*

In § 2 we will shift our concentration from B and its submodules to C.

§ 2 THE INJECTIVITY CLASS OF A MODULE

Let R be a fixed ring and C a fixed R-module. We will examine the class of all R-modules B for which, for every submodule A of B, the domain of every map $h : A \to C$ can be extended all the way to B.

DEFINITION. The *injectivity class* of an R-module C is the class $\mathrm{Inj}(C)$ of all modules B for which for each submodule A of B and each map $h : A \to C$ there is a map $k : B \to C$ for which $k \mid A = h$.

OBSERVATION 3. The injectivity class of a module is closed under quotients.

Verification: Let B/K be a quotient module of a module B in $\mathrm{Inj}(C)$. Any submodule of B/K has the form A/K for some submodule A of B containing K. To show that B/K is in $\mathrm{Inj}(C)$, consider Figure 7, where the unlabeled maps are natural and the numbered maps are constructed in numerical order as follows.

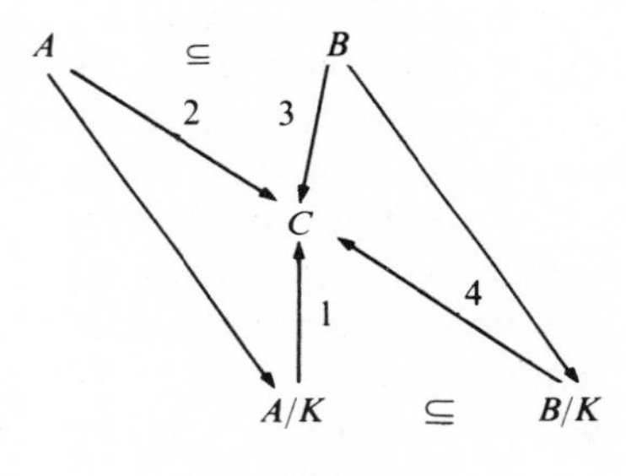

FIGURE 7

Map 1 is an arbitrarily specified map. Map 2 arises by composition. Map 3 is one of the maps whose existence is guaranteed by the condition that B be in $\mathrm{Inj}(C)$. We will show that map 3 induces map 4. If $b + K = b' + K$, then $b - b'$ is in K and we need only show that the image of $b - b'$ under map 3 is zero. But the image of K under map 3 = its image under map 2 = its image under map 1 = 0. Thus, b and b' have the same image under map 3 and map 4 is induced. That the restriction of map 4 to A/K is map 1 follows from the commutativity of the parallelogram, the commutativity of the three upper triangles, and the fact that $A \to A/K$ is an epimorphism.

OBSERVATION 4. The injectivity class of a module is closed under direct sums.

Verification: Let $\{B_i \mid i \in I\}$ be a family of modules in $\mathrm{Inj}(C)$. Let A be a submodule of $\oplus\{B_i \mid i \in I\}$ and $h : A \to C$ be an arbitrary map. Enlarge h maximally to a map $k : M \to C$. We must show that $M =$

$\oplus\{B_i \mid i \in I\}$. To do this, consider Figure 8, where j is an arbitrary element of I, map 1 arises by restriction of k, and map 2 is one of the maps whose existence is guaranteed by the condition that B_j be in Inj(C). Since k and 2 agree on their intersection, there is a map with domain $M + B_j$ that enlarges k. Since k cannot be strictly enlarged, we conclude that $B_j \subseteq M$. Since j was an arbitrary element of I, we have $\oplus B_i = M$.

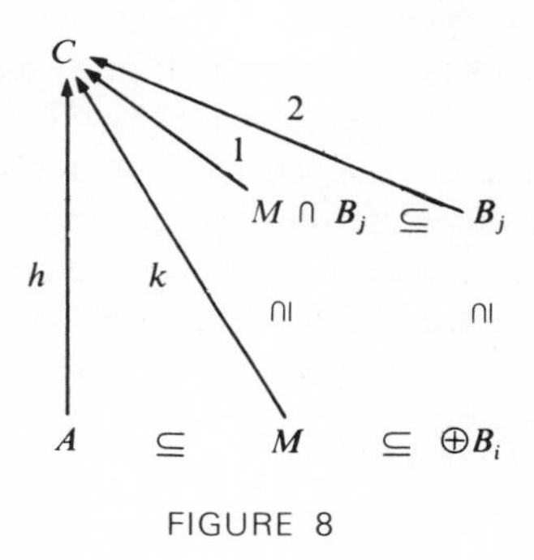

FIGURE 8

Part Two of this book focuses on those R-modules C for which every R-module lies in Inj(C).

DEFINITION. An R-module C is *injective* if for each submodule A of each R-module B and each map $h: A \to C$ there is a map $k: B \to C$ for which $k \mid A = h$.

The decision as to whether a given R-module C is or is not injective is generally made by using the following result, which is often called the *injective test lemma*.

LEMMA. An R-module C is injective if (and obviously only if) for each left ideal L and each map $h: L \to C$ there is a map $k: R \to C$ for which $k \mid L = h$.

Proof: The hypothesis asserts that R is in Inj(C). By Observation 4, all free modules are in Inj(C). By Observation 3 and Chapter 5, all R-modules are in Inj(C). Thus, C is injective.

§ 3 GENERATORS

In deriving the injective test lemma, the R-module R was used as a generator in the sense of the following definition.

DEFINITION. An R-module G is a *generator* if every R-module is an epimorphic image of a direct sum of copies of G.

We may use this concept to restate the injective test lemma in the slightly stronger form: an R-module C is injective if and only if $\mathrm{Inj}(C)$ contains a generator.

This concept of a generator has been used widely in recent years in several advanced branches of algebra. It may be interesting to locate this concept and some of its uses in Bass [1968], Freyd [1964], and MacLane [1971]. There are several equivalent definitions of a generator, which are included in the exercises. The generators that are most widely used are also projective. They are sometimes referred to by the contracted form *progenerator*. For each ring R there is an easy example of a projective generator, namely R. Further examples of generators are included in the exercises.

NOTES FOR CHAPTER 6

Sections 1 and 2 of this chapter constitute a detailed display of the elements of the proof of the injective test lemma. If you would like to see a compact proof of the lemma, I suggest page 9 of Cartan and Eilenberg [1956]. The injective test lemma is attributed to Baer [1940] by Cartan and Eilenberg.

EXERCISES

1. Prove that a submodule A of an R-module B must be a direct summand of B if A is an injective module (by using the injectivity of A to obtain a splitting map for the inclusion of A in B).
2. Prove that each direct summand of an injective module is injective. (This is proved in Chapter 8, Observation 1.)
3. a. Prove that a ring is semisimple if and only if all of its modules are injective.
   b. Prove that a ring is semisimple if and only if each of its (maximal) left ideals is injective.
4. Let R be a semisimple ring. Which R-modules are generators?
5. For which rings R is every nonzero R-module a generator?
6. Which Z-modules are generators?

7. Prove that an R-module G is a generator if and only if for every pair of R-modules A, B and every pair of distinct maps $h_1 : A \to B$, $h_2 : A \to B$, there is a map $k : G \to A$ for which $h_1 k$ and $h_2 k$ are distinct.

8. Prove that an R-module G is a generator if and only if R is isomorphic with a direct summand of a direct sum of a finite number of copies of G.

9. Prove that an R-module G is a generator if and only if R contains left ideals $L_1, \ldots, L_n$ and elements $x_1 \in L_1, \ldots, x_n \in L_n$ for which each L_i is a homomorphic image of G and $x_1 + \cdots + x_n = 1$.

10. Let R be a ring with the property that each of its left ideals is free. Prove that an R-module G is a generator if and only if G contains a direct summand isomorphic with R.

The remaining exercises are on the level of Chapter 0, but they contain information that is useful for following the diagrammatic proofs that appear throughout Part Two.

11. Let $h_1 : A \to B$, $h_2 : A \to B$, and $k : B \to C$ be maps of R-modules for which k is a monomorphism and $kh_1 = kh_2$. Show that $h_1 = h_2$.

12. Let $h : A \to B$, $k_1 : B \to C$, and $k_2 : B \to C$ be maps of R-modules for which h is an epimorphism and $k_1 h = k_2 h$. Show that $k_1 = k_2$.

13. Let $\{B_i \mid i \in I\}$ be a family of R-modules, and let $h : A \to \pi B_i$ and $k : A \to \pi B_i$ be R-homomorphisms for which $p_j h = p_j k$ for every projection $p_j : \pi B_i \to B_j$ ($j \in I$). Show that $h = k$.

CHAPTER 7

EMBEDDING MODULES IN INJECTIVE MODULES

§ 1 INTRODUCTION

It is one of the basic facts of module theory that for each ring R every R-module can be embedded in an injective R-module. The injective test lemma indicates how we should begin the embedding process. If A is an R-module that is not injective, then there is a left ideal L of R and a map $h : L \to A$ that cannot be enlarged to a map of R into A. It is an easy matter to embed A in a larger module B so that h can be enlarged to a map of R into B. This embedding process, although simple, is the most fundamental construction in this chapter and is described rather formally.

OBSERVATION 1. For A an R-module, L a left ideal, and h: $L \to A$ a map, the graph, $\text{gr}(-h)$, of the map $-h$ is a submodule of $R \oplus A$. The function $i : A \to (R \oplus A)/\text{gr}(-h)$ defined by $i(a) = (0, a) + \text{gr}(-h)$ is a monomorphism. The function $k : R \to (R \oplus A)/\text{gr}(-h)$

defined by $k(r) = (r, 0) + \mathrm{gr}(-h)$ is a homomorphism that enlarges the composite ih.

Verification: There is little here that needs comment. The maps discussed provide a diagram (Figure 9), and the last assertion of the observation is that the diagram is commutative, which we will verify. For x in L we have $k(x) = (x, 0) + \mathrm{gr}(-h) = (x, 0) - (x, -h(x)) + \mathrm{gr}(-h) = (0, h(x)) + \mathrm{gr}(-h) = ih(x)$.

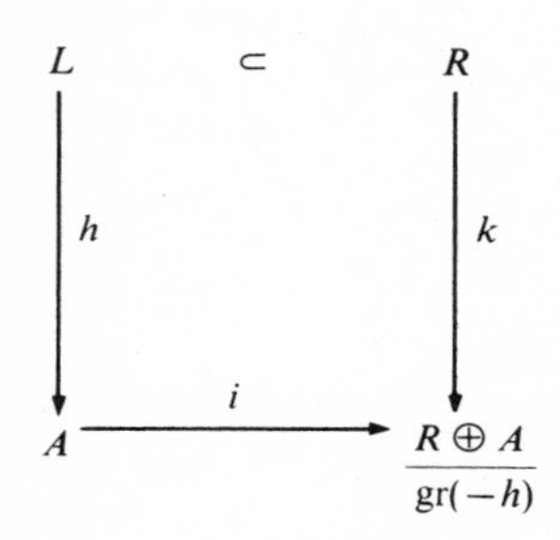

FIGURE 9

Referring to the observation, when $h: L \to A$ is given, we may identify A with its image in $B = (R \oplus A)/\mathrm{gr}(-h)$; then we have $A \subseteq B$, and h can be enlarged to a map $k: R \to B$. The new module B needn't be injective. It is merely big enough to allow the one map h to be enlarged to a map with domain R. Other maps $h_1 : L_1 \to A$ may not admit such enlargements. Moreover, for B to be injective we must be able to enlarge also maps $h_2 : L_2 \to B$. Nevertheless, it is possible to build an injective module containing A by shrewdly repeating the simple construction we have made. If we attempt to make our construction in this manner, we meet the need for infinite ordinals immediately. The approach we will take in the next section will allow us to put off working with infinite ordinals until we absolutely must.

§ 2 FOR NOETHERIAN RINGS

First, we will expand Observation 1. Let H be the set of all maps $h : L \to A$ for the various left ideals L of R. Let F be the direct sum of copies of R, one for each h in H. We will write $F = \oplus\{R_h \mid h \in H\} = \oplus R_h$. For each h, its domain L will be regarded as a submodule of R_h, and R_h will be regarded as a submodule of F. Then, for each h in H, $\mathrm{gr}(-h)$

is a submodule of $R_h \oplus A$, which is a submodule of $F \oplus A$. Let G be the submodule of $F \oplus A$ generated by $\cup\{\text{gr}(-h) \mid h \in H\}$.

OBSERVATION 2. The function $i: A \to (F \oplus A)/G$ defined by $i(a) = (0, a) + G$ is a monomorphism. For each h in H the function $k_h : R_h \oplus A \to (F \oplus A)/G$ defined by $k_h(r, a) = (r, a) + G$ is a homomorphism with kernel $\text{gr}(-h)$. For each h in H, ih can be enlarged to a map $k : R \to (F \oplus A)/G$.

Verification: We will only comment on the third assertion since the first two follow from the definitions of G and $\text{gr}(-h)$. For a fixed h in H we have a commutative diagram

$$\begin{array}{ccccc} \text{im}(i) & \subseteq & \text{im}(k_h) & \subseteq & (F \oplus A)/G \\ \uparrow i & & \wr\| & & \\ A & \subseteq & B & & \end{array}$$

where B is the module constructed by means of h in § 1. By Observation 1, h can be enlarged to a map $k' : R \to B$, and therefore ih can be enlarged to a map $k : R \to (F \oplus A)/G$.

Referring to the observation, we may identify A with its image in $C = (F \oplus A)/G$; then we have $A \subseteq C$, and each $h : L \to A$ can be enlarged to a map $k : R \to C$. The new module C needn't be injective. There may be a map $h_1 : L \to C$ that cannot be enlarged to a map with domain R. The present procedure is easy to iterate, and its iteration will yield injective modules.

To describe the iteration procedure, we will start over with fresh notation. Let A_0 be an arbitrary R-module. Let A_1 be an R-module containing A_0 with the property that each map $h : L \to A_0$ can be enlarged to a map $k : R \to A_1$. Let A_2 be an R-module containing A_1 with the property that each map $h : L \to A_1$ can be enlarged to a map $k : R \to A_2$. Continue in this way to produce a nest of R-modules $A_0 \subseteq A_1 \subseteq A_2 \subseteq \cdots \subseteq A_n \subseteq \cdots$ (n a nonnegative integer). Now define $A = \cup A_n$. Is A necessarily injective? It needn't be for arbitrary rings R. However, for one very important class of rings, such an A must be injective.

DEFINITION. A ring R is *Noetherian* if each left ideal of R is finitely generated.

Principal ideal domains are examples of Noetherian rings. For each Noetherian ring R it is known (see page 70 of Lambek [1966]) that the ring $R[x]$ of polynomials in x over R is also Noetherian. Thus, for each division ring D, the ring of polynomials in n variables $D[x_1, \ldots, x_n]$ is Noetherian. Noetherian rings take their name from Emmy Noether, whose work played a major role in shaping twentieth-century mathematical thought. (See Kimberling [1972] and the references given therein.)

The condition in Observation 3 that R be Noetherian will be eliminated in § 3. However, the Noetherian case is basically simpler, since we do not need to use infinite ordinals in giving the proof.

OBSERVATION 3. If R is Noetherian, each R-module may be embedded in an injective R-module.

Verification: We have embedded (above) the arbitrary R-module A_0 in the union $A = \cup A_n$ of the nest $\{A_n \mid n$ a nonnegative integer$\}$ of modules. We need only verify that A is injective. Let L be an arbitrary left ideal of R and $h : L \to A$ an arbitrary map of L into A. In L there must be a finite subset $\{a_1, \ldots, a_n\}$ that generates L. Each $h(a_i)$ is contained in some $A_{j(i)}$. Among the modules $A_{j(1)}, \ldots, A_{j(n)}$, the one with the greatest subscript contains all the others. Let the greatest subscript be m. Then $\mathrm{im}(h) \subseteq A_m$. The construction of A_{m+1} provided for an enlargement $k : R \to A_{m+1} (\subseteq A)$ of h. Thus, every map $h : L \to A$ can be enlarged to a map $k : R \to A$, and we conclude by the injective test lemma that A is injective.

§ 3 (Ord.) FOR ARBITRARY RINGS

What can be done when the ring R is not Noetherian? In this case we continue the nest $A_0 \subseteq A_1 \subseteq \cdots$ transfinitely. For A_ω we choose $\cup A_n$. For $A_{\omega+1}$ we choose a module containing A_ω and having the property that every map $h : L \to A_\omega$ can be enlarged to a map $k : R \to A_{\omega+1}$. In general, the definition is as follows: for each limit ordinal α we define $A_\alpha = \cup\{A_\beta \mid \beta < \alpha\}$. For each nonlimit ordinal $\alpha + 1$ we choose $A_{\alpha+1}$ to be a module containing A_α and having the property that every $h : L \to A_\alpha$ can be enlarged to a map $k : R \to A_{\alpha+1}$. We now have a nest of modules $\{A_\alpha \mid \alpha$ as ordinal$\}$. We will show that there is an ordinal λ for which A_λ is injective. Our problem now is to think of an ordinal λ for which we can imagine a proof that the corresponding module is injective. If we reexamine the Noetherian case, we will see how to choose such a λ.

In choosing our λ, we will not attempt to make a minimal choice. Let κ be the cardinal number of the ring R. Let λ be the least ordinal of cardinal $\kappa + 1$. Then λ is a limit ordinal and $A_\lambda = \cup \{A_\alpha | \alpha < \lambda\}$. Moreover, for each $\alpha < \lambda$, the cardinal of α cannot exceed κ. We can show that A_λ is injective in much the same way we verified that $A = A_\omega$ is injective when R is Noetherian. Let L be any left ideal of R and let $h : L \to A_\lambda$ be any map. For each x in L, there is an $\alpha(x) < \lambda$ such that $h(x) \in A_{\alpha(x)}$. Let $\beta = \sup\{\alpha(x) | x \in L\}$. Since the cardinal of each ordinal $\alpha(x)$ does not exceed κ and since the cardinal of L does not exceed κ, it follows that the cardinal of the ordinal β does not exceed $\kappa^2 = \kappa$. Consequently, $\beta < \lambda$ and, furthermore, $\beta + 1 < \lambda$. We have $\operatorname{im}(h) \subseteq \cup\{A_{\alpha(x)} | x \in L\} = A_\beta \subseteq A_{\beta+1} \subseteq A_\lambda$. By the choice of $A_{\beta+1}$, h can be enlarged to a map $k : R \to A_{\beta+1}$. We may regard k as a map $k : R \to A_\lambda$; as such, k is an enlargement of h. By the injective test lemma, A_λ is injective. We have now demonstrated the following observation.

OBSERVATION 4. For every ring R, each R-module can be embedded in an injective R-module.

§ 4 INJECTIVE HULLS

We have seen how an arbitrary R-module A can be embedded in an injective R-module Q. We made little effort toward economy in this embedding process; that is, the Q produced by our construction process will often be far larger than necessary, even for some modules over Noetherian rings. We will show that there is a most economical choice for the injective Q containing A and that this choice is, in a fairly strong sense, unique. This economical choice will be described by means of the following concept, which will be a valuable tool throughout our study of injective modules.

DEFINITION. A submodule A of an R-module B is *large in B* if the only submodule S of B that satisfies $A \cap S = 0$ is $S = 0$.

When A is large in B, it is also said that B is an *essential extension* of A.

DEFINITION. An injective R-module Q containing the R-module A is an *injective hull* for A if A is large in Q.

The following observation will show us where to look to find an injective hull of a module A.

OBSERVATION 5. If an R-module A is contained in an injective hull Q, then, for any injective module Q' containing A, the inclusion $A \subseteq Q'$ can be enlarged to a monomorphism of Q into Q'.

Verification: By the injectivity of Q', the inclusion $A \subseteq Q'$ can be enlarged to a map $k : Q \to Q'$. Since A is large in Q, the kernel of k must be zero.

This observation tells us that, to find an injective hull for an R-module A, we may embed A in any way we wish (say as in the preceding sections of the chapter) in an injective module Q and then search among the submodules S between A and Q. According to the observation, if A has an injective hull, at least one such intermediate submodule S must be an injective hull for A. What further properties must such an intermediate submodule have if it is to be an injective hull of A? It must contain A as a large submodule. Since it must be injective, no submodule of Q can contain S as a large submodule. When we put these two requirements together, we have the following statement. For a submodule S intermediate between A and Q to be an injective hull of A, it is necessary that S be maximal among the submodules of Q that contain A as a large submodule. Must there always be at least one intermediate submodule that satisfies this condition?

OBSERVATION 6. For each submodule A of each module B there is an intermediate submodule S that is maximal among the submodules of B that contain A as a large submodule.

Verification: Let $\mathscr{F}$ be the family of all submodules of B that contain A as a large submodule. Let $\mathscr{M}$ be a maximal nest in $\mathscr{F}$. Let S be the union of $\mathscr{M}$. Then A is large in S, and, by the maximality of $\mathscr{M}$, no submodule of B that properly contains S can contain A as a large submodule.

In our search for an injective hull for an arbitrary R-module A, our attention has now been narrowed down sufficiently. We prove that *for each ring R and each R-module A, A can be embedded in a module Q, which is an injective hull for A.*

OBSERVATION 7. Let Q be any injective R-module containing the R-module A, and let S be any submodule of Q that is maximal among the submodules of Q that contain A as a large submodule. Then S is an injective hull for A.

Verification: We need only show that S is a direct summand of Q (see Exercise 2 of Chapter 6 or Observation 1 of Chapter 8). Let M be any submodule of Q that is maximal subject to the requirement $S \cap M = 0$ (see Exercise 6). Notice that the maximality condition has two large consequences (see Exercise 7): $S \oplus M$ is large in Q, and $(S \oplus M)/M$ is large in Q/M. I will show that $Q = S \oplus M$.

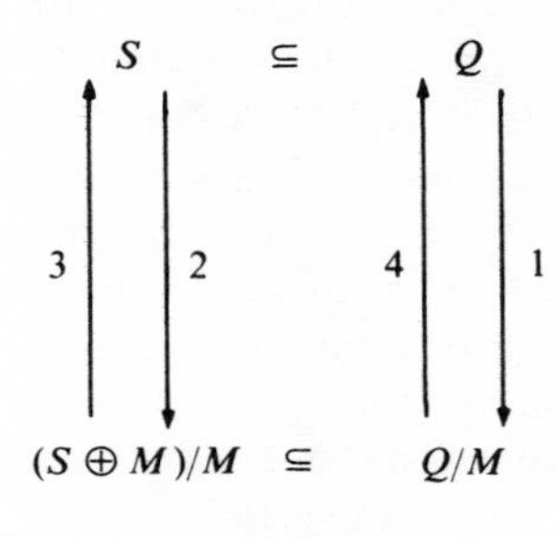

FIGURE 10

Consider the diagram shown in Figure 10, where the maps are constructed as follows. Map 1 is the natural map. Map 2 is the restriction of 1 to S with its image displayed. Since $S \cap M = 0$, map 2 is an isomorphism. Map 3 is the inverse of 2. By the injectivity of Q, there is a map 4 that enlarges 3. Now 4 is monic since its restriction to the large submodule $(S \oplus M)/M$ is 3, which is monic (see Exercise 3). Since $(S \oplus M)/M$ is large in Q/M, the image of $(S \oplus M)/M$ under 4 is large in the image of 4 (see Exercise 3). Thus, S is large in im(4). Since A is large in S and S is large in im(4), we have A large in im(4) (see Exercise 4). Since S is maximal subject to containing A as a large submodule, im(4) $= S$. Since 4 is monic and has the same image as 3, we conclude that $(S \oplus M)/M = Q/M$ and $S \oplus M = Q$ as desired.

The degree of uniqueness of the injective hull of a module was already implicit in Observation 5 and is spelled out in the following observation.

OBSERVATION 8. If Q and Q' are injective hulls of A, then the identity map on A can be enlarged to an isomorphism between Q and Q'.

Verification: By the injectivity of Q, the inclusion $A \subseteq Q$ can be enlarged to a map $k : Q' \to Q$. Since A is large in Q', k is a monomorphism. Since $\operatorname{im}(k) \cong Q'$ is an injective submodule of Q that contains A, we have $Q = \operatorname{im}(k) \oplus S$ from which we conclude $S = 0$ since A is large in Q.

§ 5 DIVISIBLE MODULES

For certain rings it seems natural to approach injectivity from a point of view that is more element oriented as opposed to the general map-oriented viewpoint. This approach will lead us to the attempt to divide elements of an R-module by elements of R. An element a of an R-module A is *divisible* by an element r of R if there is an x in A for which $a = rx$. There is a very blunt limitation on divisibility that is most conveniently described in terms of the following concept, widely used in ring theory.

DEFINITION. The *annihilator* $\operatorname{Ann}(a)$ of an element a of an R-module A is the subset $\{r \in R \mid ra = 0\}$ of R.

OBSERVATION 9. If an element a of an R-module A is divisible by r, then $\operatorname{Ann}(a) \supseteq \operatorname{Ann}(r)$.

Because of this observation, we will never consider the possibility of dividing an element a by an element r unless $\operatorname{Ann}(a) \supseteq \operatorname{Ann}(r)$. The following definition is made with this in mind.

DEFINITION. An R-module A is *divisible* if for each r in R and a in A that satisfy $\operatorname{Ann}(r) \subseteq \operatorname{Ann}(a)$ there is an x in A for which $a = rx$.

We begin to see the relationship between divisibility and injectivity with the following observation.

OBSERVATION 10. An injective module is necessarily divisible.

Verification: Let A be an injective R-module, and let r and a be elements of R and A that satisfy $\operatorname{Ann}(r) \subseteq \operatorname{Ann}(a)$. A map $h : Rr \to A$ is defined by $h(r'r) = r'a$ as follows. Suppose $r'r = r''r$. Then $(r' - r'') \in \operatorname{Ann}(r) \subseteq \operatorname{Ann}(a)$ and consequently $r'a = r''a$. Thus, h is a function and also a map. By the injectivity of A, there is an enlargement $k : R \to A$ of h. Then $rk(1) = k(r) = h(r) = a$, confirming the divisibility of A.

For an arbitrary ring R the divisibility of a module needn't imply injectivity. There is, however, a significant class of rings for which divisibility is equivalent to injectivity.

OBSERVATION 11. If R is a ring that has the property that each left ideal is cyclic, then an R-module A is injective if (and only if) it is divisible.

Verification: Let L be an arbitrary left ideal of R and $h: L \to A$ an arbitrary map. Let r be an element of R for which $L = Rr$. Since $\mathrm{Ann}(r) \subseteq \mathrm{Ann}(h(r))$, there is, by the divisibility of A, an x in A for which $h(r) = rx$. The map $k : R \to A$ defined by $k(r') = r'x$ enlarges h since $k(r'r) = r'rx = r'h(r) = h(r'r)$.

Some common examples of rings for which every left ideal is cyclic are semisimple rings, principal ideal domains, and proper quotient rings of principal domains (such as Z/Zn for a positive integer n).

We will restrict our attention for the duration of this section to modules over a principal ideal domain R. Since R has no proper zero divisors, the restriction placed on division by the annihilator inclusion condition evaporates. It merely prevents division of a nonzero-module element by the zero element of R.

In the study of the structure of injective modules (Chapter 9), one of the critical topics will be the determination of injective hulls of certain cyclic modules. In the present circumstances we have a rather convenient way of describing some of these injective hulls. By Observation 7, *a uniform divisible* R*-module must be an injective hull of each of its nonzero submodules.* Let Q be the module (field) of fractions of the principal ideal domain R. Then Q is certainly a divisible module, and it is also uniform since for any two nonzero elements r_1/r_2, r_3/r_4 of Q the cyclic submodules generated by r_1/r_2 and r_3/r_4 both contain the nonzero element r_1r_3. Thus, Q is a divisible hull of each of its nonzero submodules. In particular, Q is an injective hull for R.

For each prime p in R we will construct one additional uniform injective R-module. (In Chapter 9 we will find that we have here constructed, up to isomorphism, every uniform injective R-module.) Notice that for each positive integer n, the cyclic R-module R/Rp^{n+1} contains a cyclic submodule isomorphic to R/Rp^n. Specifically, $p + Rp^{n+1}$ generates such a submodule. This observation allows us to construct a nest of R-modules as follows. Let $A_1 = R/Rp$. There exists an R-module $A_2 \cong R/Rp^2$ satisfying $A_1 \subset A_2$. There exists an R-module $A_3 \cong R/Rp^3$ satisfying $A_2 \subset A_3$. By continuing in this way, produce a nest of R-modules, $A_1 \subset A_2 \subset \cdots \subset A_i \subset \cdots$ (i a positive integer), for which $A_i \cong R/Rp^i$. Let A be the union of this nest of R-modules. We will show that A is a divisible uniform R-module.

Let a be a nonzero element of A. Let n be the least positive integer for which a is in A_n. For any nonzero element r in R we have $r = p^k q$, where k is a nonnegative integer and q is an element of R which is relatively prime to p. To verify that A is divisible, we need to solve $a = rx$. For an element y in A_{n+k} we have $a = p^k y$. We can divide y by q by means of an elementary Euclidean style of argument: now $p^m y = 0$ for $m = n + k$, and, since q is relatively prime to p, there are elements u, v in R for which $1 = up^m + vq$. Then $y = up^m y + vqy = q(vy)$. Finally, $a = p^k y = p^k q(vy) = r(vy)$, and we conclude that A is divisible. The cyclic submodule generated by a is $A_n \supseteq A_1$; consequently, since a is an arbitrary nonzero element of A, A is uniform.

Since A is uniform and divisible (that is, injective), A is an injective hull for each of the cyclic submodules $A_i \cong R/Rp^i$. *We will adopt $R(p^\infty)$ as our standard notation for the isomorphism type of this module A.*

NOTES FOR CHAPTER 7

The ability of an arbitrary module to be embedded in an injective is a major feature of module theory. This fact was exploited systematically by Cartan and Eilenberg in [1956], where they attributed the discovery of this embeddability to R. Baer [1940]. The equally fundamental theory of the injective hull is due to Eckmann and Schopf [1953]. For a slicker proof of the embeddability, consult Rotman [1970]. I prefer the proof presented in this chapter because each step is dictated by the injective test lemma. For an important use of injective hulls not taken up here, see the discussion of rings of quotients given in Lambek [1966].

Although it was tricky to embed an arbitrary module in an injective (§§ 1, 2, 3), once the job was done we found that there was an essentially unique, economical way to do it (§ 4). Although it was trivial to represent an arbitrary module as the quotient of a projective (even free) module (Chapter 5), there is, in general, no way to do this that is the best possible way in a sense that parallels the economy of the injective hulls among injective embeddings. For clarification of my meaning, examine the concept of a projective cover introduced by Bass [1960] or see the attractive list of exercises in Lambek [1966, p. 93].

EXERCISES

1. Prove that an R-module A is injective if and only if whenever A appears as a submodule of an R-module B it is a direct summand of B.
2. a. Show that a ring R is Noetherian if and only if it contains no infinite nest of distinct left ideals of the form $L_1 \subset L_2 \subset \cdots \subset L_i \subset \cdots$ (i a positive integer).
   b. Show that a ring R is Noetherian if and only if it contains no infinite sequence of elements $r_1, r_2, \ldots, r_i, \ldots$ (i a positive integer) for which the left ideals $Rr_1 \subset Rr_1 + Rr_2 \subset \cdots \subset Rr_1 + \cdots + Rr_i \subset \cdots$ are all distinct.
3. Let $h : B \to C$ be a homomorphism of R-modules, and let A be a large submodule of B.
   a. Show that it is not possible in general to conclude that $h(A)$ is large in $h(B)$.
   b. Show that if $h \mid A$ is monic then h is monic.
   c. Show that if h is monic then $h(A)$ is large in $h(B)$.
4. Let A be a large submodule of B and B be a large submodule of C. Show that A is large in C.
5. Let S be an independent subset of a module B, and let A be the submodule generated by S. Prove that A is a large submodule of B if and only if S is a maximal independent subset of B.
6. Let A be a submodule of B. Prove that B contains a submodule M for which $A \cap M = 0$ and $A \cap N \neq 0$ for any submodule N properly containing M. (We say that such a submodule M is maximal subject to the condition $A \cap M = 0$.)
7. Let S be a submodule of A, and let M be a submodule of A that is maximal subject to $S \cap M = 0$.
   a. Prove that $S \oplus M$ is large in A.
   b. Prove that $(S \oplus M)/M$ is large in A/M.
8. Prove that an indecomposable injective module must be uniform.
9. Let S be a submodule of an injective module A, and let M be a submodule of A that is maximal subject to $S \cap M = 0$. Prove that M is injective.
10. Show that a quotient module of a divisible module need not be divisible.
11. Let R be a principal ideal domain. Show that the class of injective R-modules has the following closure properties:
    a. Direct summands of injective R-modules are injective.
    b. The direct product of any family of injective R-modules is injective.
    c. Quotient modules of injective R-modules are injective.

d. The direct sum of any family of injective R-modules is injective.
e. If A is an R-module that is the union of a nest of injective submodules, then A is injective.

12. Let R be a principal ideal domain, and let F be the field of fractions of R. Suppose that p is a prime of R. We define $F(p) = \{r/p^n \mid r \in R, n \text{ is a nonnegative integer}\}$ and observe that $F(p)$ is an R-submodule of F that contains R. For any nonnegative n, the inclusions $R \subseteq F(p) \subseteq F$ yield $R/Rp^n \subseteq F(p)/Rp^n$. Show that $F(p)/Rp^n$ is an injective hull for R/Rp^n when n is a positive integer.

13. Let p be a prime integer. For each positive integer n, let $C(p^n)$ be the set of all those complex numbers that are p^nth roots of 1.
a. Verify that each $C(p^n)$ is a subgroup of the multiplicative group of complex numbers of absolute value 1.
b. Verify that $C(p^n)$ is cyclic of order p^n and that $C(p) \subset C(p^2) \subset \cdots \subset C(p^i) \subset \cdots$ (i a positive integer).
c. Let $C(p^\infty) = \cup \{C(p^n) \mid n \text{ a positive integer}\}$. Show that $C(p^\infty)$ is an injective hull of each $C(p^n)$. (Here we are regarding $C(p^\infty)$ and its subgroups as Z-modules, where the notation is multiplicative rather than additive.)

14. Let Z_n be the ring of integers mod n and regard Z_n as a module over itself in the usual way.
a. Show that if n is a power of a prime then Z_n is divisible and therefore injective (as a Z_n-module).
b. Show that Z_n is divisible (and therefore injective) even when n is not a power of a prime.

15. Let R be a principal ideal domain, and let I be a nonzero ideal of R. Regard the ring R/I as a module over itself in the usual way. Show that R/I is divisible and therefore injective (as an R/I-module).

16. Let R be a ring with the property that every injective R-module is free.
a. Show that every free R-module must be injective.
b. Show that the following classes of R-modules coincide: projective, injective, free.

CHAPTER 8

INJECTIVE MODULES AS A CLASS

§ 1 GENERAL CLOSURE PROPERTIES

Before we begin our structural investigation of injective modules, we will determine the closure properties of the class of injective R-modules for various rings R.

For an arbitrary ring R the class of injective modules has only two closure properties of significance.

OBSERVATION 1. Each direct summand D of each injective R-module C is injective.

Verification: Let A be a submodule of an R-module B, and let $h : A \to D$ be an arbitrary map. Consider the diagram shown in Figure 11, where the numbered maps are constructed as follows. Map 1 is one of the enlargements of h; the existence of map 1 is guaranteed by the injectivity of C. Map 2 is one of the enlargements of the identity map on D; the

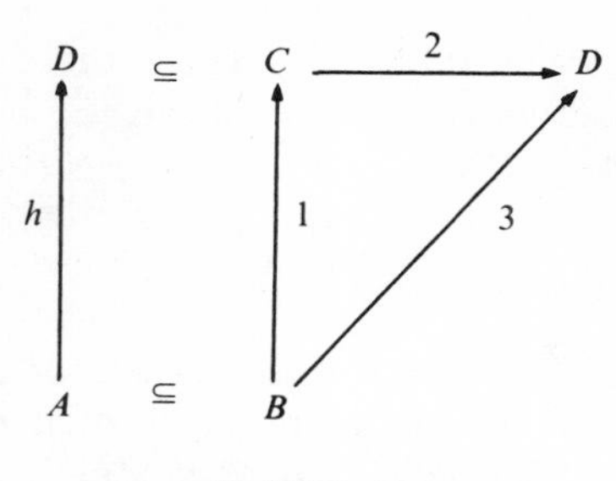

FIGURE 11

existence of map 2 is guaranteed by the summand feature of D. Map 3 is the composite of 1 and 2. That the restriction of 3 to A is h follows from the fact that 1 and 2 are enlargements.

OBSERVATION 2. The direct product $\pi\{C_i \mid i \in I\}$ of a family $\{C_i \mid i \in I\}$ of R-modules is injective if (and only if) each member of the family is injective.

Verification: Suppose that each C_i $(i \in I)$ is injective, A is a submodule of an R-module B, and $h : A \to \pi C_i$ is an arbitrary map. Consider the diagram shown in Figure 12, where j is an arbitrary element of I and the

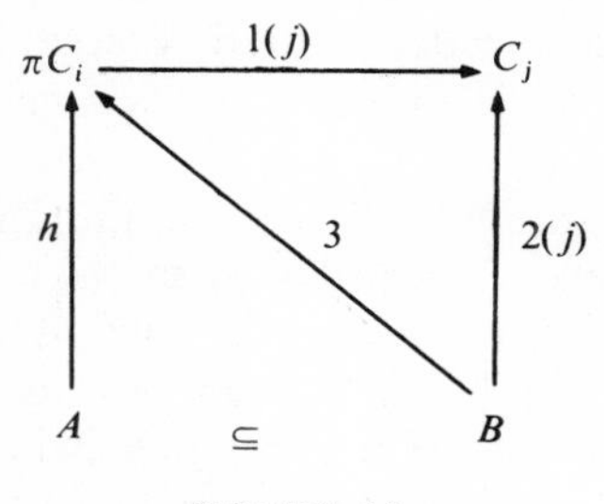

FIGURE 12

numbered maps are constructed as follows. Map $1(j)$ is the cannonical projection of the product onto its coordinate C_j. Map $2(j)$ is one of the enlargements of the composite of h and $1(j)$; the existence of map $2(j)$ is guaranteed by the injectivity of C_j. Map 3 is the product of the family of maps $\{2(j) \mid j \in I\}$. That the restriction of 3 to A is h follows from the fact that, when either is composed with any $1(j)$, the result agrees with $2(j)$.

Notice the important consequence: the direct sum of a *finite* family of injective modules is always injective.

§ 2 COGENERATORS

We have the proper tools to construct, for each ring R, modules that have properties that are diagrammatically dual to those of generators. Recall that in Chapter 6 I defined an R-module G to be a generator if every R-module is an epimorphic image of a direct sum of copies of G.

DEFINITION. An R-module C is a *cogenerator* if every R-module is isomorphic to a submodule of a direct product of copies of C.

As for generators, one always has the simple example of a generator that is also projective: R. In a few special cases a cogenerator is also readily available. For example, if R is semisimple, then R is a cogenerator that is also injective. For an arbitrary ring it is not as evident that a cogenerator exists.

OBSERVATION 3. For each ring R there is an injective cogenerator.

Verification: Let $\{L_i \mid i \in I\}$ be the set of all left ideals of R. For each $i \in I$ let C_i be an injective R-module containing R/L_i. The direct product $C = \pi\{C_i \mid i \in I\}$ is injective, and we will verify that it is a cogenerator.

Let A be an R-module. Then A contains a maximal independent set S. (Independent sets were discussed in § 3 of Chapter 2.) The submodule A' generated by S is the direct sum $\oplus\{Rs \mid s \in S\}$ of the cyclic modules $Rs (s \in S)$. For each s in S there is a left ideal L_j of R and an isomorphism $h_s : Rs \to R/L_j$. Composing h_s with the embedding $R/L_j \subseteq C_j \subseteq C$ gives for each s in S an embedding $k_s : Rs \to C$. The family of embeddings $\{k_s \mid s \in S\}$ provides an embedding $k : \oplus Rs \to D$, where D is the direct *sum* of the various copies of C, one copy for each s in S. Let E be the corresponding direct *product* in which D is contained. We may regard k as having range E and write $k : \oplus Rs \to E$. Consider the diagram shown in Figure 13, where k' is one of the enlargements of k, the existence of which is guaranteed by the injectivity of E. Since E is a direct product of copies of C and A is an arbitrary R-module, we need only show that $\ker(k') = 0$.

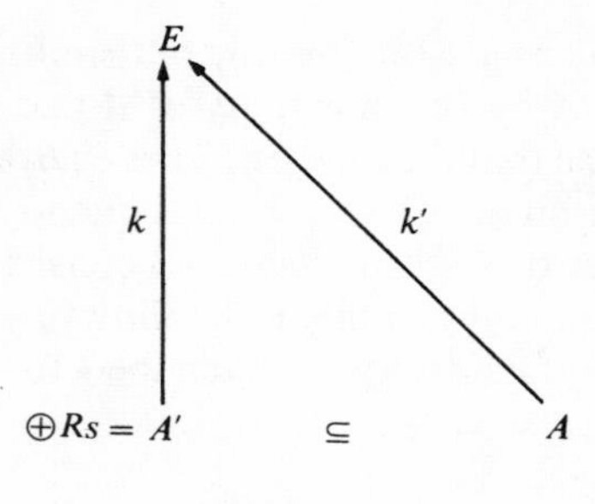

FIGURE 13

Since k is a monomorphism, $A' \cap \ker(k') = 0$ and for any t in $\ker(k')$ we have a direct sum $A' \oplus Rt$. If $t \neq 0$, then it follows from the directness of this sum that $S \cup \{t\}$ is an independent set. Since S is a maximal independent set, we conclude that $t = 0$ and $\ker(k') = 0$ as desired.

The injective test lemma has as a corollary the following: an R-module A is injective if, for each projective R-module B and each submodule K of B, every map $h : K \to A$ can be enlarged to a map with domain B. We will prove the dual of this latter statement.

OBSERVATION 4. An R-module P is projective if, for each injective R-module A and each submodule K of A, every map $h : P \to A/K$ can be lifted through the natural map of A onto A/K.

Verification: Let B be an arbitrary R-module, K any submodule of B, and $h : P \to B/K$ a map. Embed B as a submodule in an injective R-module A. We have a diagram (Figure 14) where the horizontal maps are natural and the numbered maps arise as follows. Map 1 consists of

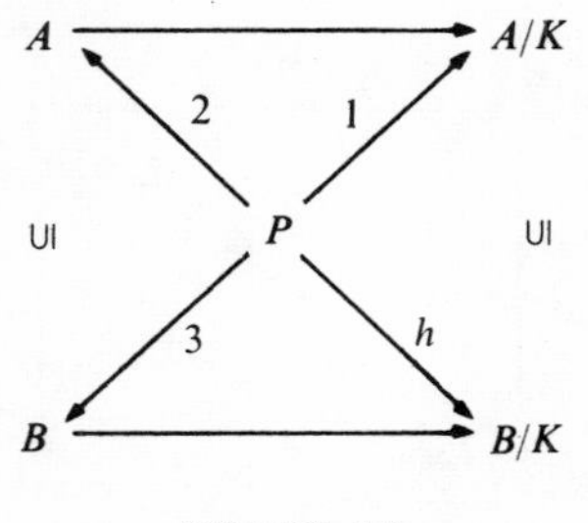

FIGURE 14

the same ordered pairs as h. Map 2 is one of the liftings of 1; the existence of map 2 is guaranteed by the injectivity of A and the hypothesis of the observation. It happens that the image of 2 lies in B: if a is in the image of 2, then $a + K$ is in the image of 1 and consequently $a + K = b + K$ for some b in B. It follows that a is in B. Map 3 consists of the same ordered pairs as 2. That 3 is actually a lifting of h follows by appealing first to the commutativity of the outside square and then to the commutativity of the three upper triangles.

This observation will prove to be a convenient tool in the next section. Cogenerators will not play an important role in this book, but they may be expected to play a significant role in future developments. The concept of a cogenerator is used, for example, in Osofsky [1966] and in Rutter [1971]. You may wish to examine discussions of cogenerators in MacLane [1971] also.

§ 3 HEREDITARY RINGS (AGAIN)

When we ask for those rings for which every quotient module of an injective module is injective, we are led back to a familiar class.

PROPOSITION 4. A ring R has the property that each quotient module of each injective module is injective if and only if R is hereditary.

Proof: Suppose R is hereditary and K is a submodule of an injective R-module A. We must show that A/K is injective. Let L be an arbitrary left ideal of R and $h : L \to A/K$ a map. Consider the diagram shown in Figure 15, where the numbered maps are constructed in numerical order as follows. Map 1 is the natural map. Map 2 is one of the liftings of h through 1; the existence of map 2 is guaranteed by the projectivity of L.

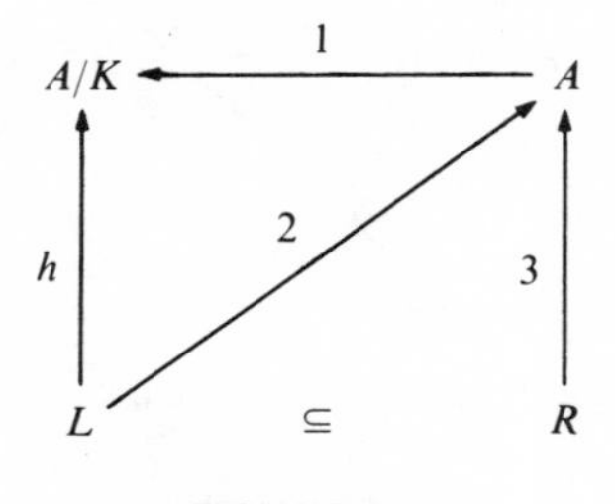

FIGURE 15

Map 3 is one of the enlargements of 2; the existence of map 3 is guaranteed by the injectivity of A. That h is the restriction to L of the composite of 3 and 1 is a consequence of the commutativity of the two triangles.

Suppose that R has the property that quotient modules of injective modules are always injective. Let A be an arbitrary injective R-module, K an arbitrary submodule of A, L a left ideal, and $h : L \to A/K$ a map. By Observation 4, to prove that L is projective, we need only show that h can be lifted through the natural map of A onto A/K. Consider the diagram shown in Figure 16, where the numbered maps are con-

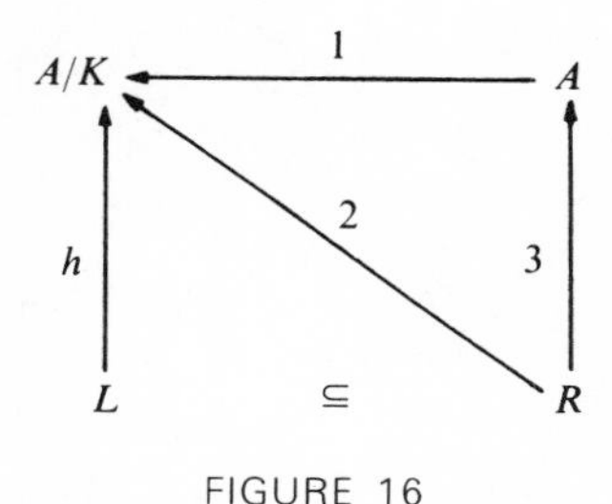

FIGURE 16

structed in numerical order as follows. Map 1 is the natural map. Map 2 is one of the enlargements of h; the existence of map 2 is guaranteed by the injectivity of A/K. Map 3 is one of the liftings of 2 through 1; the existence of map 3 is guaranteed by the projectivity of R. That the restriction of 3 to L is a lifting of h through 1 is a consequence of the commutativity of the two triangles.

For which rings R is every R-module isomorphic with a quotient module of an injective module? The answer to this question is the class of quasi-Frobenius rings that will be discussed briefly in § 5 of Chapter 9. A proof will not be given in this book, but the problem will be discussed in the notes to Chapter 9.

§ 4 NOETHERIAN RINGS

For which rings is the class of injective modules closed under direct sums? The answer to this question will take us back to the class of Noetherian rings introduced in § 2 of Chapter 7.

OBSERVATION 5. If R is Noetherian and $\{C_i \mid i \in I\}$ is an arbitrary family of injective R-modules, then $\oplus\{C_i \mid i \in I\}$ is injective.

Verification: The injective test lemma is a powerful tool here. Let L be a left ideal of R and $h: L \to \oplus C_i$ a map. Consider the diagram shown in Figure 17, where F and the numbered maps are described as follows. Since R is Noetherian, L is finitely generated. Then im(h) is also

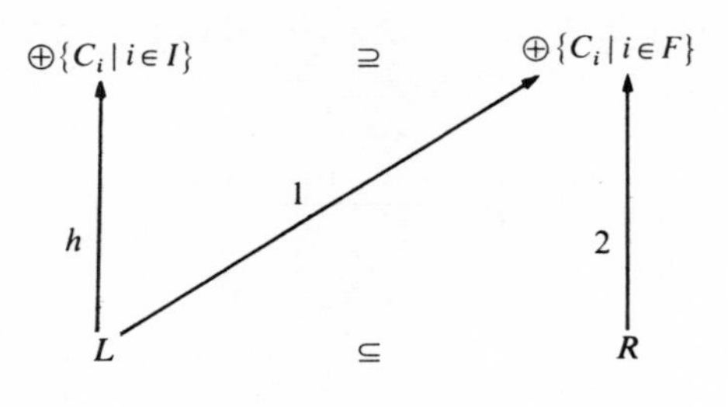

FIGURE 17

finitely generated, and, by the nature of a direct sum, there is a finite subset F of I for which im(h) $\subseteq \oplus\{C_i \mid i \in F\}$. Map 1 is the function consisting of the same ordered pairs as h. Since $\oplus\{C_i \mid i \in F\}$ is a direct *product* of injective modules, it is injective; as a result, there exists an enlargement, map 2, of 1 as required.

OBSERVATION 6. If a ring R has the property that the direct sum of every countable family, $\{C_i \mid i \in N\}$, of injective modules is injective, then each submodule of each finitely generated R-module is finitely generated.

Verification: Let R have the property mentioned in the observation, and let A be an arbitrary submodule of a finitely generated R-module B. We will verify the observation by assuming that A is not finitely generated and producing a contradiction: since A is not finitely generated, elements $a_1, a_2, \ldots, a_i, \ldots$ (i a positive integer) can be chosen in succession from A so that $Ra_1 \subset Ra_1 + Ra_2 \subset \cdots \subset Ra_1 + Ra_2 + \cdots + Ra_i \subset \cdots$ is a strictly ascending nest of submodules of A. For each i, let A_i denote $Ra_1 + Ra_2 + \cdots + Ra_i$, and let $A' = \cup A_i$. For each x in A' there is an n such that x is in A_j for $j \geq n$. Thus, $h(x) = (x + A_1, x + A_2, \ldots, x + A_i, \ldots)$ defines a map $h: A' \to \oplus\{A'/A_i \mid i \text{ a positive integer}\}$. For each i let $A'/A_i \subseteq Q_i$ be an embedding of A'/A_i in an injective module Q_i. We may regard h as a map $h: A' \to \oplus Q_i$. Since R satisfies the hypothesis of the observation, $\oplus Q_i$ is injective and h may be enlarged to a map $k : B \to \oplus Q_i$. Let p_j be the jth projection $p_j : \oplus Q_i \to Q_j$. We have two contradictory statements concerning $\mathrm{im}(h)$: Since B is finitely generated, $p_j k(B) = 0$ for all but finitely many j. On the other hand, for each j, $p_j k(B) \supseteq p_j k(A') = A'/A_j \neq 0$.

PROPOSITION 5. A ring R has the property that the direct sum of every countable family, $\{C_i \mid i \in N\}$, of injective modules is injective if and only if R is Noetherian.

Proof: If R is Noetherian, the required closure property follows from the stronger one stated in Observation 5. If R has the property that the direct sum of every countable family, $\{C_i \mid i \in N\}$, of injective modules is injective, then each left ideal of R must be finitely generated by Observation 6.

The injectives over a Noetherian ring have another simple closure property.

OBSERVATION 7. If R is Noetherian and C is an R-module that is the union of a nest $\{C_i \mid i \in I\}$ of injective submodules, then C is injective.

Verification: Let L be a left ideal of R and $h : L \to C$ a map. Since L is finitely generated and the C_i $(i \in L)$ form a nest, there is a $j \in I$ for which $\mathrm{im}(h) \subseteq C_j$. By the injectivity of C_j the map h can be enlarged to a map $k : R \to C_j \subseteq C$.

PROPOSITION 6. A ring R has the property that the union of each nest of injective modules of the form $C_1 \subseteq C_2 \subseteq \cdots \subseteq C_i \subseteq \cdots (i \in N)$ is injective if and only if R is Noetherian.

Proof: If R is Noetherian, the required closure property follows from the stronger one stated in Observation 7. Suppose that R has the closure property stated in the proposition. Any countable family of R-modules can be indexed with positive integers. Thus, let $\{A_i \mid i \in N\}$ be any countable family of injective R-modules. Then $\oplus\{A_i \mid i \in N\}$ is the union of the nest of submodules $A_1 \subseteq A_1 \oplus A_2 \subseteq \cdots \subseteq A_1 \oplus \cdots \oplus A_i \subseteq \ldots$. Each member of the nest is injective, and consequently $\oplus A_i$ is injective. By the previous proposition, R must be Noetherian.

NOTES FOR CHAPTER 8

In summary, the class of injective R-modules is closed under:

1. summands for all R,
2. direct products for all R,
3. quotients if and only if R is hereditary,
4. direct sums if and only if R is Noetherian,
5. unions of nests if and only if R is Noetherian, and
6. submodules if and only if R is semisimple.

The observation that the closure of the class of injectives under direct sums (or unions of nests) necessitates the ring being Noetherian is attributed by Rotman [1970] to H. Bass. The remaining closure properties appear in Cartan and Eilenberg [1956].

With the closure properties in mind, the process of embedding an arbitrary R-module in an injective can sometimes be reduced to embedding R itself in an injective R-module. This is explained in Exercise 2.

EXERCISES

1. Give new proofs of Corollaries 1 and 2 of Chapter 3 by imitating the proofs of Observations 1 and 2 of the present chapter. (Your proofs will be "dual" to the proofs of Observations 1 and 2 in the sense that they will be based on diagrams that are similar to the diagrams appearing in the proofs of the observations but with arrows reversed and monics and epics interchanged.)

2. Let R be a ring, and let R be contained in an R-module C. Let A be an arbitrary R-module.
   a. Show that A is isomorphic with a submodule of $\oplus\{C_i \mid i \in I\}/S$ for some family $\{C_i \mid i \in I\}$ of copies of C and some submodule S of $\oplus C_i$ (Chapter 5 is useful here).
   b. Suppose further that R is hereditary and that C is injective. Show that A can be embedded in an injective module of the form $\pi\{C_i \mid i \in I\}/S$.
   c. Suppose now that R is also Noetherian. Show that A can be embedded in an injective module of the form $\oplus C_i/S$.

3. Let R be a ring, and let A be an arbitrary R-module. Suppose that $\{C_i \mid i \in I\}$ is a collection of injective submodules of A, and let C be the submodule of A generated by $\{C_i \mid i \in I\}$.
   a. Show that if R is hereditary and I is finite then C must be injective.
   b. Show that if R is hereditary and Noetherian then C must be injective regardless of the cardinality of I.

4. Let R be a Noetherian ring, and let A be an arbitrary R-module.
   a. Show that A contains a maximal injective submodule C. Can you determine whether C must be unique?

b. Show that $A = B \oplus C$, where C is injective and B contains no nonzero injective submodule.
c. Show that if R is not only Noetherian but also hereditary then A contains a unique maximal injective submodule.

5. Let R be a ring. Prove that R has the property that each submodule of each of its finitely generated modules is finitely generated if and only if R is Noetherian. (This is proved in Chapter 10 where it appears as Proposition 7.)

6. a. Let A be an R-module that contains a finite independent set $\{m_1, \ldots, m_k\}$, which generates a large submodule of A. For each $i\,(1 \leq i \leq k)$ let A_i be an injective hull for the cyclic submodule Rm_i of A. Show that A is isomorphic with a submodule of $\oplus\{A_i \mid 1 \leq i \leq k\}$.
b. Let R be a Noetherian ring, A an R-module, and M a maximal independent subset of A. For each m in M let A_m be an injective hull for the cyclic submodule Rm of A. Show that A is isomorphic with a submodule of $\oplus\{A_m \mid m \in M\}$.

CHAPTER 9

INJECTIVE MODULES: STRUCTURE

§ 1 INTRODUCTION

We have come to the climax of Part Two. Three tools will be fundamental for our structural investigations: (1) the concept of a uniform module introduced in Chapter 1, (2) our knowledge of injective hulls, and (3) some special features of modules over Noetherian rings. A reasonably good insight into the structure of uniform injective modules can be developed. For arbitrary injective modules the structure will only be clarified when the ring is Noetherian.

Twice before we have seen Noetherian rings behave especially well in relation to injectivity. The construction of injective hulls is appreciably easier when the ring is Noetherian. For Noetherian rings, the class of injectives is closed under direct sums. This closure property and the good behavior to be exhibited in § 4 with respect to uniformity are the keys to the determination of the structure of the injective modules over Noetherian rings.

I will give no structural results for arbitrary injective modules over rings that are not Noetherian. The device that we consistently use for analyzing the structure of modules is the method of direct sum decomposition. A few moments of consideration will suggest that a penetrating analysis of the structure of the injective modules by this means could be expected only for rings for which the class of injective modules is closed under direct sums. As we have seen, these rings are precisely the Noetherian rings. To determine the structure of the injectives over non-Noetherian rings, other methods will be needed.

§ 2 UNIFORM INJECTIVE MODULES

It is possible to reduce the problem of determining the structure of the uniform injective modules over an arbitrary ring to considerations involving the left ideals of the ring. This possibility is suggested by the principle that each nonzero element of a uniform module is interrelated with the overall structure of the module. In the present situation this principle takes the following form.

OBSERVATION 1. For each nonzero element c of a nonzero uniform injective R-module H, Rc is a uniform cyclic module and H is an injective hull of Rc.

Verification: Since H is uniform, Rc is uniform and large in H. Since H is injective, it must be a hull of Rc.

The observation suggests a program for cataloging the uniform injectives over R. The first step should be to catalog the uniform cyclic R-modules. Since each cyclic R-module is isomorphic to a quotient R/L for some left ideal L, this step merely requires determining which left ideals produce uniform quotients. The second step should be to determine when two uniform modules R/L and R/L' have isomorphic injective hulls. We begin by answering the following question. For which left ideals L of R is R/L uniform?

DEFINITION. A left ideal L of a ring R is *meet-irreducible* if $L \neq L' \cap L''$ for any left ideals L' and L'' that properly contain L.

OBSERVATION 2. For a left ideal L of R, R/L is uniform if and only if L is meet-irreducible.

Verification: If $L = R$, the observation is immediate. For the remainder of the verification, we will assume that L is proper. Suppose R/L is uniform and L', L'' are left ideals that properly contain L. Since R/L is uniform, the intersection of its (nonzero) submodules L'/L and L''/L is not zero. This means $L' \cap L'' \neq L$.

Suppose L is meet-irreducible. Any pair of nonzero submodules of R/L are of the form L'/L, L''/L, where L', L'' are left ideals of R that properly contain L. By the meet-irreducibility of L, $L' \cap L'' \neq L$ and $(L'/L) \cap (L''/L) = (L' \cap L'')/L$, which is not zero.

When do uniform cyclic R-modules have isomorphic injective hulls?

OBSERVATION 3. Injective hulls, H and H', of nonzero uniform cyclic R-modules C and C' are isomorphic if and only if there are nonzero submodules S and S' of C and C' for which $S \cong S'$.

Verification: Suppose that $h : H \to H'$ is an isomorphism. Since H' is uniform, the submodule S' defined by $S' = C' \cap h(C)$ is nonzero. For $S = h^{-1}(S')$ we have $S \cong S'$.

Suppose that $h : S \to S'$ is an isomorphism where S and S' are nonzero submodules of C and C' (or even of H and H'). Let $k : H \to H'$ be one of the enlargements of h whose existence is guaranteed by the injectivity of H'. Since S is large in C, it is also large in H, and from $\ker(k) \cap S = 0$ we conclude that k is a monomorphism. Then $k(H)$ is injective, and from the uniformity of H' we conclude that k is an isomorphism.

We can now give an explicit description of our cataloging procedure. Let $\mathscr{L}$ be the set of all meet-irreducible left ideals of R. For L, L' in $\mathscr{L}$ we write $L \sim L'$ if R/L and R/L' contain isomorphic nonzero submodules. From Observation 3 we know that $\sim$ is an equivalence relation in $\mathscr{L}$ and that the injective hulls of R/L and R/L' are isomorphic if and only if $L \sim L'$. *A complete catalog, without repetitions, of the uniform injective modules over an arbitrary ring R may be obtained as follows: from each $\sim$ equivalence of meet-irreducible left ideals of R, choose one member L and then form the injective hull of R/L.*

Our cataloging procedure will be illustrated for a very special type of ring in the next section.

In Matlis [1958] the uniform injectives over commutative Noetherian rings were classified in a satisfactory manner. This classification is outlined in Exercise 11 through 17.

§ 3 FOR PRINCIPAL IDEAL DOMAINS

We will illustrate the program described in the preceding section by cataloging the uniform injective modules over a principal ideal domain R. In order to discuss the ideals of R, we first choose a representative set P of primes for R. Thus, for each prime p in R there is a *unique* p' in P such that p divides p'. Each ideal is of the form Ra for some a in R. When is Ra meet-irreducible? If $a = 0$, then $Ra = 0$ is meet-irreducible by the absence of zero-divisors in R. If a is a unit, then $Ra = R$ is meet-irreducible (but of no interest to us). Now suppose that a is neither zero nor a unit. We may assume that $a = p_1^{e_1} \ldots p_n^{e_n}$, where n and the exponents are positive integers and each p_i is in P (see Exercise 1). If $n > 1$, then Ra is not meet-irreducible because $Ra = Rp_1^{e_1} \cap Rp_2^{e_1} \cap \cdots \cap Rp_n^{e_n}$. We know that each nonzero proper meet-irreducible ideal is of the form Rp^e for some p in P and some positive exponent e. Now $Rp^e \sim Rp$ since R/Rp^e contains the submodule Rp^{e-1}/Rp^e, which is isomorphic with R/Rp. Thus, every nonzero meet-irreducible ideal of R is equivalent to one of $\{Rp \mid p \in P\}$. Is it possible that $Rp \sim Rq$ for distinct elements p, q of P? If $Rp \sim Rq$ then since R/Rp and R/Rq are simple they would have to be isomorphic. Such an isomorphism is not possible because if $1 + Rp$ corresponded under such an isomorphism with $r + Rq (\neq 0)$ we would have $p(1 + Rp) = 0$ corresponding with $p(r + Rq) = pr + Rq \neq 0$.

We have observed that $\{0\} \cup \{Rp \mid p \in P\}$ is a complete set of proper inequivalent meet-irreducible (left) ideals of R. We need only describe an injective hull for each of the corresponding quotient modules: $\{R\} \cup \{R/Rp \mid p \in P\}$. We have already described hulls for these modules in § 5 of Chapter 7. On recalling our results there, we can conclude our description of the uniform injective R-modules as follows.

For a principal ideal domain R, each nonzero uniform injective R-module is isomorphic to one and only one of the modules in $\{F\} \cup \{R(p^\infty) \mid p \in P\}$, where F is the module (field) of fractions of R and P is a representative set of primes in R.

In the next section we will see that every injective module over a principal ideal ring R is a direct sum of these uniform injective modules.

We have seen that when R is a principal ideal domain, each nontrivial meet-irreducible ideal L is $\sim$ equivalent to precisely one ideal of the form Rp (p a prime). Exercise 17a indicates the sense in which this situation is typical for arbitrary commutative Noetherian rings.

§ 4 FOR NOETHERIAN RINGS

In this section the classification of the injective modules over a Noetherian ring will be reduced to the program of classifying its uniform injectives as described in § 2. This reduction is accomplished by using two powerful tools that are available for modules over these rings. One tool is the closure of the class of injective modules under direct sums, which we saw in Observation 5 of Chapter 8. The second tool is the fact that every module over a Noetherian ring contains a large semi-uniform submodule. This latter fact will be established in the course of the next three observations.

OBSERVATION 4. Each nonzero cyclic module over a Noetherian ring contains a nonzero uniform submodule.

Verification: Let L be a proper left ideal of an arbitrary ring R, and suppose that $A = R/L$ contains no nonzero uniform submodule. We will reserve symbols with A as base for *nonzero* submodules of A. Since A is not uniform we have $A \supseteq A_1 \oplus A'_1$. Since A'_1 is not uniform, we have $A'_1 \supseteq A_2 \oplus A'_2$. Continuing in this way, we have an inclusion $A'_i \supseteq A_{i+1} \oplus A'_{i+1}$ for every positive integer i. Then $R/L = A \supseteq \oplus\{A_i \mid i \text{ is a positive integer}\}$. We have $\oplus A_i = K/L$ for some left ideal K satisfying $L \subseteq K \subseteq R$. Since the (strictly infinite) direct sum $\oplus A_i$ cannot be finitely generated, neither can the left ideal K. Thus, R is not Noetherian.

OBSERVATION 5. Each nonzero module A over a Noetherian ring R contains a nonzero uniform cyclic submodule.

Verification: Let B be a nonzero cyclic submodule of A. By Observation 4, B contains a nonzero uniform submodule C. Let D be any nonzero cyclic submodule of C. Since any submodule of a uniform module is uniform, D is a nonzero uniform cyclic submodule of A.

OBSERVATION 6. Let A be a module over a Noetherian ring R. The submodule B generated by a maximal uniformly independent set M is necessarily large in A.

Verification: Suppose that B is not large in A. Then there is a nonzero submodule C of A that yields a direct sum $B \oplus C$. By Observation 5, C contains an element $c \neq 0$ for which Rc is uniform. Then $M \cup \{c\}$ is a uniformly independent subset that is strictly larger than M. Since M is a maximal uniformly independent subset, we have arrived at a contradiction. We conclude that B is large in A.

THEOREM 3. Each injective module over each Noetherian ring is semiuniform.

Proof: Let A be an injective module over a Noetherian ring R. Let M be a maximal uniformly independent subset of A. For each m in M let A_m be an injective hull of Rm in A. Since each A_m contains the uniform module Rm as a large submodule, A_m is uniform. From the directness of $\oplus\{Rm \mid m \in M\}$ we infer the directness of $\oplus\{A_m \mid m \in M\}$. (See Exercise 5.) From the maximality of M and Observation 6 we conclude that $\oplus Rm$ is large in A. Then $\oplus A_m$ is certainly also large in A, and, since R is Noetherian, it is also injective. Since $\oplus A_m$ is an injective large submodule of A, it must coincide with A. Thus, $A = \oplus\{A_m \mid m \in M\}$, where each $A_m (m \in M)$ is uniform.

A principal ideal domain is necessarily Noetherian; in § 3 we cataloged the uniform injective modules over such rings. Theorem 3, together with this catalog, yields the following corollary.

COROLLARY. For a principal ideal domain R, an R-module is injective if and only if it is isomorphic to the direct sum of a family of copies of modules selected from the set $\{F\} \cup \{R(p^\infty) \mid p \in P\}$, where F is the module (field) of fractions of R and P is a representative set of primes in R.

This corollary may be regarded as a special case of the result of Matlis, which appears as Exercise 17b.

§ 5 PROJECTIVE INJECTIVES AND QUASI-FROBENIUS RINGS

The basic theme of Part One was projectivity. We were able to do quite a bit with the general problem of determining the structure of projective modules. Let us return to this theme and ask the following question: What is the structure of a module that is both injective and projective? Since we have just discussed injectives over principal ideal domains, the question may seem peculiar. For principal ideal domains, the only module that is a projective injective is the zero-module. If we think back to semisimple rings, however, we see another picture altogether. Every module over a semisimple ring is both projective and injective.

Our experience with injective modules suggests that we begin by asking for the structure of the uniform projective injective modules.

OBSERVATION 7. For an arbitrary ring R the uniform projective injective R-modules are precisely the indecomposable injective principal R-modules.

Verification: Let A be a uniform projective injective R-module. Since A is uniform and projective, it is isomorphic to a left ideal of R by Observation 4 of Chapter 1. Since A is injective, this ideal is a summand of R. Since A is uniform, this ideal is indecomposable. Let A be an indecomposable injective summand of R. Since A is injective and indecomposable, it is also uniform. As a summand of R, A is certainly also projective.

For Noetherian rings this observation extends to a structure theorem for arbitrary projective injective modules.

THEOREM 4. The projective injective modules over a Noetherian ring R are, up to isomorphism, precisely the direct sums of indecomposable injective principal R-modules.

Proof: Since R is Noetherian, every direct sum of injective principal R-modules is injective and of course also projective. Since R is Noetherian, each projective injective module is a direct sum of uniform injective modules, and each of these summands must also be projective. By Observation 7, each of these summands is therefore an indecomposable injective principal R-module.

It is possible to display the uniform injective left ideals of a Noetherian ring R in a rather neat manner. Since the class of injective modules over a Noetherian ring is closed under the formation of unions of nests, R contains a maximal injective left ideal A. (See also Exercise 2a, Chapter 7.) Then $R = A \oplus B$ for some left ideal B, and B *contains no nonzero injective submodule.* Since A is injective, it is semiuniform, and we have

$A = \oplus\{A_i \mid i \in I\}$. Then $R = (\oplus A_i) \oplus B$. Since R is cyclic, it is only finitely decomposable. Thus, I is finite and we may assume $I = \{1, \dots, n\}$ for some n. We have $R = A_1 \oplus \cdots \oplus A_n \oplus B$, where the A_i are uniform projective injective left ideals. Any uniform projective injective R-module C is isomorphic to a left ideal and must by its uniformity be isomorphic with a submodule of some A_i. (Recall Observation 4 of Chapter 1.) Since the image of C in A_i is injective, we must have $C \cong A_i$ (or $C \cong 0$) by the uniformity of A_i. Thus, $R = A_1 \oplus \cdots \oplus A_n \oplus B$ provides a complete display (possibly with repetitions) of the isomorphism types of uniform projective injective R-modules. Notice the following corollary: *up to isomorphism there are only finitely many uniform projective injective modules over a Noetherian ring.*

There is a widely studied class of rings, the quasi-Frobenius rings, that have been characterized in many ways. A close examination of these rings would carry us in a direction tangential to the basic program of this book. However, the structure of the projective and(?) injective modules over quasi-Frobenius rings is known and is available to us instantly if we will set down the following highly unorthodox definition of a quasi-Frobenius ring. In spite of the peculiarity of the definition, it is actually equivalent to the various other definitions in common use (such as the one given in Curtis and Reiner [1962]).

DEFINITION. A ring R is *quasi-Frobenius* if the class of projective R-modules coincides with the class of injective R-modules.

Notice that semisimple rings are quasi-Frobenius since each module over a semisimple ring is both projective and injective. It is not very difficult to verify that the proper quotient rings of principal ideal domains are always quasi-Frobenius. One of the stimuli to the development of the theory of quasi-Frobenius rings has been the following fact, which I will not prove: if the characteristic of a field F is a prime that divides the order of a finite group G, then the group ring $F(G)$ is not semisimple but it is quasi-Frobenius.

In order to derive the structure of the projective = injective modules over quasi-Frobenius rings from Theorem 4, we need only two simple facts: A quasi-Frobenius ring R is necessarily Noetherian since its class of injectives is closed under direct sums. For R quasi-Frobenius, R is injective, and therefore all the principal R-modules are also injective.

COROLLARY. The projective = injective modules over a quasi-Frobenius ring are, up to isomorphism, precisely the semiprincipal modules.

Notice that as a special case of our earlier discussion we have for each quasi-Frobenius ring R a decomposition $R = A_1 \oplus \cdots \oplus A_n$, where each A_i is a uniform projective = injective R-module and every uniform projective = injective R-module is isomorphic with at least one $A_i (1 \leq i \leq n)$. The situation here strongly resembles the special case in which R is semisimple.

NOTES FOR CHAPTER 9

I attribute Chapter 9 as a whole to Eben Matlis [1958]. Attention was drawn to the class of modules that are simultaneously projective and injective by J. P. Jans [1959]. Our unusual definition of quasi-Frobenius rings can be connected to more orthodox definitions by means of the paper by Carl Faith and Elbert Walker [1967] and the paper by Faith [1966].

From Matlis [1958] you can learn the following uniqueness result that is an important adjunct to Theorem 3: if R is a Noetherian ring and $\{A_i \mid i \in I\}$ and $\{B_j \mid j \in J\}$ are families of uniform injective R-modules for which $\oplus A_i \cong \oplus B_j$, then there is a bijective function $f: I \to J$ for which $A_i \cong B_{f(i)}$ for all i in I.

From Faith [1966] and Faith & Walker [1967] the answers to the questions raised on pages 63 and 93 can be given. Specifically, the following assertion can be justified. For a ring R the following seven assertions are equivalent:

1. Each R-module is isomorphic to a submodule of a free module.
2. Each R-module is isomorphic to a submodule of a projective module.
3. All injective R-modules are projective.
4. R is quasi-Frobenius.
5. All projective R-modules are injective.
6. All free R-modules are injective.
7. Each R-module is isomorphic to a quotient module of an injective module.

The equivalence of 3 and 4 is established in Faith and Walker [1967] and the equivalence of 4 and 5 is established in Faith [1966]. The remainder of the assertion is covered by Exercise 9 below. Rutter [1969], [197?] has given some further characterizations of quasi-Frobenius rings including a characterization closely related to condition 1 of the list above.

EXERCISES

1. Let R be a principal ideal domain, and let P be a representative set of primes in R. Show that if an element a of R is neither zero nor a unit then $Ra = Rp_1^{e_1} \cdots p_n^{e_n}$, where n and the exponents are positive integers and each p_i is in P.

2. Let R be a principal ideal domain, p a prime in R, and e a positive integer. Show that Rp^e is meet-irreducible.

3. Let A be a large submodule of an R-module B. Show that if A is uniform then B is also uniform.

4. Let A be an R-module containing submodules B and C for which $B \cap C = 0$. Show that if B is large in a submodule D of A and C is large in a submodule E of A then $D \cap E = 0$.

5. Let A be an R-module containing a family $\{C_i \mid i \in I\}$ of submodules for which the submodule generated by the family is the *direct* sum $\oplus\{C_i \mid i \in I\}$. Suppose that, for each $i \in I$, D_i is a submodule of A that contains C_i as a large submodule. Show that the submodule generated by the family $\{D_i \mid i \in I\}$ is the direct sum $\oplus\{D_i \mid i \in I\}$.

6. Give a thorough description of the structure of the injective modules over each of the following rings:
   a. The ring consisting of those rational numbers that are expressible with odd denominators.
   b. The ring consisting of those rational numbers that are expressible with denominators relatively prime to 6.
   c. The ring consisting of those rational functions in the indeterminant x over the field F (quotients of polynomials in x over F) that are expressible with denominators having nonzero constant term.
   d. The ring consisting of those rational functions in x over F that are expressible with denominators relatively prime to $x^2 + x$.

7. Let R be the ring of integers mod p^n, where p is a prime and n is a positive integer.
   a. With Exercise 14 of Chapter 7 in mind, classify the uniform injective R-modules by carrying out the procedure of § 2.
   b. Apply Theorem 3 to show that the injective R-modules are precisely the free R-modules.
   c. Conclude that all projective R-modules are free.
   d. Observe that R is a quasi-Frobenius ring.
   e. Generalize a through d to the case in which R is a quotient ring of an arbitrary principal ideal domain modulo the ideal generated by a nonzero power of a prime element.

8. Let R be a quasi-Frobenius ring.
   a. Show that every R-module is isomorphic to a submodule of a projective (free) R-module.
   b. Show that every R-module is isomorphic to a quotient module of an injective R-module.

9. Verify the following assertions concerning a ring R:
   a. Each R-module is isomorphic to a submodule of a projective (free) R-module if and only if each injective R-module is projective.
   b. Each R-module is isomorphic to a quotient module of an injective R-module if and only if each projective (free) R-module is injective.

10. Show that if a ring is both hereditary and quasi-Frobenius then it must be semisimple.

Let C be a commutative Noetherian ring. Exercises 11 through 17 will deal with C and its modules. This sequence of exercises culminates in the classification, due to Matlis [1958], of the injective modules over commutative Noetherian rings.

11. An ideal P of C is *prime* if for each $x,y \in C$ we have $xy \in P$ only when either $x \in P$ or $y \in P$.
    a. Which ideals of the ring Z of integers are prime?
    b. Show that the zero ideal is prime if and only if C is an integral domain.
    c. Show that a prime ideal must be meet-irreducible.

12. Let A be a C-module and let x be an element of A.
    a. Show that Ann(x) is an ideal of C. (Recall that Ann(x) is $\{c \in C \mid cx = 0\}$ and is called the annihilator of x.)
    b. Show that the cyclic submodule Cx of A is isomorphic with the quotient module $C/\text{Ann}(x)$.
    c. Suppose U and V are large submodules of A, that every nonzero element of U has the ideal P as its annihilator, and that every nonzero element of V has the ideal P' as its annihilator. Show that if $A \neq 0$ then $P = P'$.

13. Let P be an ideal of C. Show that P is a prime ideal if and only if each nonzero element of C/P has P as its annihilator.

14. Let $\mathscr{A}$ be the family which consists of those ideals of C that are annihilators of nonzero elements of A; that is, $\mathscr{A} = \{\text{Ann}(x) \mid 0 \neq x \in A\}$. Use the Noetherian property of C and Exercise 2a of Chapter 7 to show that there is a maximal member of $\mathscr{A}$; that is, show that there is an $a \in A$ for which Ann(a) is not properly contained in Ann(x) for any nonzero x in A.

15. Let A be a C-module and let a be an element of A for which Ann(a) is maximal in $\mathscr{A}$.
    a. Use the commutativity of C and the maximality of Ann(a) in $\mathscr{A}$ to show that the annihilator of each nonzero element of Ca is precisely Ann(a).
    b. Show that Ann(a) is a prime ideal.

16. Let A be a uniform injective C-module and let a be an element of A for which Ann(a) is maximal in $\mathscr{A}$.
    a. Show that A is isomorphic with an injective hull of $C/\text{Ann}(a)$.
    b. Let P be any prime ideal of C for which A is isomorphic with an injective hull of C/P. Show that $P = \text{Ann}(a)$.

17. Let C be a commutative Noetherian ring. Verify the following assertions:
    a. For each uniform injective C-module A there is a unique prime ideal P of C for which A is isomorphic with an injective hull of C/P.
    b. A C-module is injective if and only if it is isomorphic with the direct sum of a family of copies of modules selected from the set $\{H(P) \mid P \text{ a prime ideal of } C\}$ where for each P, $H(P)$ is an injective hull of C/P.

PART THREE

COUNTABILITY

CHAPTER 10

FINITELY GENERATED MODULES

§ 1 INTRODUCTION

We will begin our considerations of finitely generated modules by asking the following question. For an arbitrary ring R, what closure properties does the class of finitely generated R-modules have? Any quotient module of a finitely generated module must be finitely generated. Consequently the class of finitely generated R-modules is closed under quotients and summands. The direct sum of a family of nonzero finitely generated modules will be finitely generated if and only if the family is finite. In Chapter 11 countably infinite direct sums of finitely generated modules will be examined and used. Rings differ with regard to the property of submodules of finitely generated modules being necessarily finitely generated.

PROPOSITION 7. A ring R has the property that each submodule of each finitely generated R-module is finitely generated if and only if R is Noetherian.

Proof: If each submodule of each finitely generated R-module is finitely generated, then each left ideal of R must be finitely generated, and so R must be Noetherian. Suppose now that R is Noetherian. Then we know by Observation 5 of Chapter 8 that the direct sum of every family of injective R-modules is injective. It then follows by Observation 6 of Chapter 8 that each submodule of each finitely generated R-module must be finitely generated.

The preceding proof of Proposition 7 allowed us to review the powerful machinery of injectivity that was developed in Part Two. In Exercises 1 through 3, I have outlined a traditional proof of this proposition that uses only elementary concepts. For one particular class of Noetherian rings, the structure of the finitely generated modules has long been known. In the next section we will prove this structure theorem by a method that continues our review of the theory of injective modules.

§ 2 FOR PRINCIPAL IDEAL DOMAINS

Let R be a principal ideal domain, and let A be an arbitrary nonzero finitely generated R-module. We are going to embark on a discussion that will show that *every such A must be the direct sum of a finite family of cyclic submodules each of which is isomorphic either with R or with R/Rp^n for some prime p of R and some positive integer n.*

Since R is Noetherian, the previous section has provided us with the following tools to be used in our discussion: A can contain no submodule that is the direct sum of an infinite family of nonzero submodules, and, in particular, every independent subset of A is finite. According to the following argument, it will be sufficient to show that *every such A has a direct summand of one of the specified types.* If we assume that every nonzero finitely generated R-module contains a direct summand of the desired sort, then we can decompose successively, with nonzero summands C_i of the desired type, to produce $A = C_1 \oplus B_1 = C_1 \oplus C_2 \oplus B_2 = \ldots$. This process must terminate in an equality $A = C_1 \oplus C_2 \oplus \cdots \oplus C_n$ for some positive integer n since otherwise A would contain a submodule $\oplus\{C_i \mid i \text{ a positive integer}\}$, where each C_i is nonzero. Thus, we proceed to show that A has a nonzero summand isomorphic either to R or to an R/Rp^n.

Let $a_1, \ldots, a_k$ be a maximal uniformly independent subset of A. For each i $(1 \le i \le k)$ let $Ra_i \subset H_i$ be an embedding of the cyclic submodule Ra_i of A in an injective hull. The inclusion $\oplus Ra_i \subset \oplus H_i$ may be enlarged to a map $k : A \to \oplus H_i$ by the injectivity of $\oplus H_i$. Since R is Noetherian and $\{a_i\}$ is a maximal uniformly independent subset of A, $\oplus Ra_i$ is large in A. Since $\ker(k) \cap \oplus Ra_i = 0$, it follows that k is a monomorphism. Consequently from this point on we may regard A *as a submodule of* $\oplus\{H_i \mid 1 \leq i \leq k\}$, *where each* H_i *is the injective hull of a nonzero uniform cyclic* R*-module.* In § 3 of Chapter 9 we found that the nonzero uniform cyclic R-modules are precisely the modules isomorphic with R or an R/Rp^n. In § 5 of Chapter 7 we determined that the hulls of these modules are F and $R(p^\infty)$, where F is the module (field) of fractions of R and $R(p^\infty)$ was constructed as the union of a certain nest of cyclic R-modules. Now let us consider the nature of the images, C_i, of A under the projections into the various H_i $(1 \leq i \leq k)$. Each C_i is a nonzero finitely generated submodule of H_i. But each nonzero finitely generated submodule of F is isomorphic to R (see Exercise 4), and each finitely generated submodule of an $R(p^\infty)$ is isomorphic to R/Rp^m for some nonnegative integer m (see Exercise 5). Consequently from this point on we may *regard* A *as a submodule of* $\oplus\{C_i \mid 1 \leq i \leq k\}$, *where each* C_i *is either* R *or* R/Rp^m *for some prime* p *in* R *and some positive integer* m *and where, moreover, the projection of* A *into each* C_i *is an epimorphism.*

We have $A \subseteq C_1 \oplus \cdots \oplus C_k$, and by rearranging the C_i we may assume either of the following cases: (1) $C_1 = R$, or (2) $C_1 = R/Rp^m$ and there is no C_i for which either $C_i = R$ or $C_i = R/Rp^{m+t}$ for any positive integer t. In case 1 the kernel of the projection of A into C_1 is $A \cap B$, where $B = C_2 \oplus \cdots \oplus C_k$. The image of this projection is $C_1 = R$, which is projective. Consequently $A = C \oplus (A \cap B)$ for a submodule $C \cong C_1 = R$ of A. In case 2 let q be an element of R that is not divisible by p but for which $qC_i = 0$ for all C_i not of the form R/Rp^t for any positive integer t. We will use the following elementary fact. Since q is not divisible by the prime p, there are elements u, v in R such that $1 = uq + vp^m$. Now choose an element $c = (x_1, \ldots, x_k)$ in A for which x_1 generates C_1. Then the first coordinate, x_1, of C_1 may be expressed as a multiple of $qx_1 : x_1 = 1x_1 = uqx_1 + vp^m x_1 = uqx_1$. Now let C be the cyclic submodule of A generated by $qc = (qx_1, \ldots, qx_k)$, and let $B = C_2 \oplus \cdots \oplus C_k$. Then $C \cap B = 0$ because any element r of R that satisfies $rqx_1 = 0$ also satisfies $rqc = (0, 0, \ldots, 0)$. It follows that the projection of C into C_1 is a monomorphism, and it must also be an epimorphism since $x_1 = uqx_1$ is the image of uqc. Thus, $C \cong C_1 = R/Rp^m$. To complete the discussion of case 2, we need only verify that $A = C + (A \cap B)$. To do this, observe that for any $a = (y_1, \ldots, y_k)$ in A there is an r in R for which $y_1 = rx_1 = ruqx_1$, and consequently $a = ruqc + (a - ruqc)$ is in $C + (A \cap B)$. Reuniting the two cases, we have $A = C \oplus (A \cap B)$, where C is isomorphic either to R or to R/Rp^m for some prime p and some positive integer m.

Our discussion has now demonstrated the desired result, which is restated formally in the following theorem.

THEOREM 5. Each finitely generated module over a principal ideal domain R is the direct sum of a finite number of submodules each of which is isomorphic to R or R/Rp^n for some prime p and some nonnegative integer n.

NOTES FOR CHAPTER 10

The proposition and theorem that constitute this chapter are very old and are proved in many texts, even in some introductory texts in modern algebra. For this reason, I have felt free to give the unusual proofs that are likely to be new to some readers. Although I like the proofs myself, I do not mind if you regard them as a form of mathematical humor. As many readers will have noticed, I have chosen to suppress the traditional concepts of torsion modules, torsion-free modules, and chain conditions in favor of a radical emphasis on projectivity, injectivity, and cardinality restrictions as the basic concepts of structure theory for modules over arbitrary rings. This is perhaps my deeper reason for giving the unusual proof of Theorem 5. For traditional proofs of this theorem, I suggest those given in Hu [1965], Rotman [1965], and Schreier and Sperner [1959]. These references also contain clear discussions of the important question of the degree of uniqueness of decompositions of the type described in Theorem 5.

EXERCISES

1. Let A be a submodule of an R-module B. Suppose that G is a subset of A that generates A and that H is a subset of B for which $\{h + A \mid h \in H\}$ generates B/A.
   a. Show that $G \cup H$ generates B.
   b. Show that if both A and B/A are finitely generated then so is B.
2. For an arbitrary ring R, an R-module A is said to be *Noetherian* if A and each of its submodules is finitely generated.
   a. Show that quotient modules of Noetherian modules are Noetherian.
   b. Show that an R-module B must be Noetherian if it contains a submodule A for which both A and B/A are Noetherian.
3. Let R be a Noetherian ring.
   a. Use Exercise 2b and finite induction to show that each free R-module with a finite basis is a Noetherian module.

b. Use Chapter 5 and Exercise 2a to show that each finitely generated R-module is Noetherian—that is, that submodules of finitely generated modules over Noetherian rings are finitely generated.

4. Let R be a principal ideal domain, and let F be its field of fractions. Regard F as an R-module.
   a. Show that each nonzero cyclic submodule of F is isomorphic with R.
   b. Show that, for elements a/b and c/d of F $(a, b, c, d \in R)$, $R(a/b) + R(c/d) = R(g/bd)$, where g is the greatest common divisor of the elements ad and bc of R.
   c. Show that it follows from part a, part b, and a finite-induction argument that every finitely generated nonzero submodule of F is isomorphic with R.

5. Let p be a prime element in a principal ideal domain R. Recall that $R(p^\infty)$ is the union of a nest of cyclic submodules each of which is isomorphic to one of the modules R/Rp^n for a nonnegative integer n.
   a. Show that the cyclic submodule of $R(p^\infty)$ generated by an element a of $R(p^\infty)$ is the least member of this nest that contains a.
   b. Show that the submodule of $R(p^\infty)$ generated by a finite set of elements of $R(p^\infty)$ is cyclic (and may be described as the least member of this nest that contains each element of the finite set).

6. Let R be a principal ideal domain that has (essentially) only one prime; that is, each prime of R divides each other prime of R. (The ring of rational numbers expressible with odd denominators is such a ring.) Reread § 2, and find those points in the discussion that can be simplified for such rings.

CHAPTER 11

COUNTABLY GENERATED MODULES

§ 1 INTRODUCTION

What are the closure properties of the class of countably generated modules? Since a homomorphic image of a countably generated module must be countably generated, the class of countably generated modules is closed under quotients and also summands. The direct sum of a family of nonzero countably generated modules will be countably generated if and only if the family is at most countably infinite. Direct sums of arbitrary families of countably generated modules will be studied in Chapter 12, and the results will be applied in Chapter 13. Rings differ with regard to the property of submodules of countably generated modules being necessarily countably generated.

PROPOSITION 8. A ring R has the property that each submodule of each countably generated R-module is countably generated if and only if each left ideal of R is countably generated.

Since we will not be using this proposition, I will omit its proof (see Exercise 1). Probably the simplest countably generated modules are the direct sums of countable families of finitely generated modules. We will study these modules in the next section.

§ 2 DIRECT SUMS OF FINITELY GENERATED MODULES

The countably generated modules that are direct sums of finitely generated modules are easily characterized.

OBSERVATION 1. A countably generated R-module A is the direct sum of finitely generated submodules if and only if each finitely generated submodule (alternately, each finite subset) of A is contained in a finitely generated summand.

Verification: The necessity of the containment condition is clear, so we will begin by supposing that the containment condition holds for the R-module A. If A is finitely generated, no verification is needed; so suppose $a_1, a_2, \ldots, a_i, \ldots$ (i a positive integer) is a sequence of elements that generates A and that no finite subsequence generates A. Let A_1 be a finitely generated summand of A that contains a_1, and let $A = A_1 \oplus C_1$. Let B_2 be a finitely generated summand of A that contains A_1 and a_2. Let $A = B_2 \oplus C_2$. Since $A_1 \subseteq B_2$ and A_1 is a summand of A, A_1 is also a summand of B_2. Let $B_2 = A_1 \oplus A_2$. Then also $A = A_1 \oplus A_2 \oplus C_2$, and $a_1, a_2 \in A_1 \oplus A_2$. This process can be continued. Thus, there is a finitely generated summand B_3 containing $A_1 \oplus A_2$ and a_3. This yields decompositions $A = B_3 \oplus C_3$, $B_3 = A_1 \oplus A_2 \oplus A_3$, and $A = A_1 \oplus A_2 \oplus A_3 \oplus C_3$. Continuation yields a submodule $\oplus\{A_i \mid i \text{ a positive integer}\}$ of A. For each of the generators a_j, we have $a_j \in A_1 \oplus \cdots \oplus A_j$, and consequently $A = \oplus A_i$. Finally, each A_i is finitely generated since it is a homomorphic image (direct summand) of the finitely generated module B_i.

If we recast the previous observation only slightly, so that it will refer only to singleton subsets rather than finite subsets, we will produce a more effective tool.

OBSERVATION 2. A countably generated R-module A is the direct sum of finitely generated submodules if, for each summand B of A, each element of B is contained in a finitely generated direct summand of B.

Verification: The proof is similar to the proof of Observation 1 with some slight changes. We make the decomposition $A = A_1 \oplus C_1$ as before. The decomposition $A = A_1 \oplus A_2 \oplus C_2$ will be arrived at as follows. Let $a_2 = x + y$ with x in A_1 and y in C_1. Now choose A_2 to be a finitely generated summand of C_1 containing y. Letting $C_1 = A_2 \oplus C_2$, we have $A = A_1 \oplus A_2 \oplus C_2$ and $a_1, a_2 \in A_1 \oplus A_2$. Continuing in this fashion will yield $A = \oplus A_i$.

Observation 2 yields a powerful tool for the investigation of projective modules. The power of this tool will be further enhanced once we have the theorem of Kaplansky, which appears in the next chapter. The tool referred to is expressed in the following lemma.

LEMMA. A ring R has the property that each of its countably generated projective modules is the direct sum of finitely generated submodules if and only if each element of each countably generated projective R-module is contained in a finitely generated direct summand.

Proof: The necessity of the containment condition is clear. The sufficiency is obtainable from Observation 2.

In the next section we will use this lemma in the process of determining the countably generated projective modules over one more class of rings.

§ 3 SEMIHEREDITARY RINGS

Chapter 4 introduced the broadest class of rings for which we could reduce the structure problem for submodules of free modules to the study of left ideals. This was the class of hereditary rings. The hereditary rings belong to a larger class for which the structure problem for arbitrary submodules of free modules has not been reduced to the study of left ideals but for which the structure problem for finitely generated submodules of free modules has been so reduced. This larger class is defined as follows.

DEFINITION. A ring R is *semihereditary* if each finitely generated left ideal is projective.

Although we cannot handle arbitrary submodules of free modules over semihereditary rings, those that are finitely generated are very easily handled by means of the techniques of Chapter 4.

THEOREM 6. Each finitely generated submodule of each free module over a semihereditary ring R is isomorphic to a direct sum of finitely generated left ideals of R.

Proof: Notice first that a finitely generated submodule of a free module is always contained in a finitely generated free submodule. Thus, we begin by assuming that A is a finitely generated submodule of a free R-module $R_1 \oplus \cdots \oplus R_n$. If $n = 1$, then A is a finitely generated left ideal of $R = R_1$. Assume that the theorem holds for all finitely generated submodules of free modules having bases of $n - 1$ elements. Let p be the projection of A into R_1. Then $\text{im}(p)$ is a finitely generated left ideal of R and is therefore projective. Consequently, $A = A_1 \oplus \ker(p)$, where $A_1 \cong \text{im}(p)$. Since $\ker(p) \subseteq R_2 \oplus \cdots \oplus R_n$, $\ker(p)$ is isomorphic to a direct sum of finitely generated left ideals by the induction hypothesis. Thus, A is isomorphic to a direct sum of finitely generated left ideals, and by finite induction the proof is complete.

This theorem has five corollaries. The first four are immediate, and the fifth uses Observation 4 of Chapter 4. The second and third corollaries are propositions that characterize semihereditary rings.

COROLLARY 1. A finitely generated projective module over a semihereditary ring is isomorphic to a direct sum of finitely generated left ideals.

COROLLARY 2. A ring is semihereditary if and only if each finitely generated submodule of each of its free modules is projective.

COROLLARY 3. A ring is semihereditary if and only if each finitely generated submodule of each of its projective modules is projective.

COROLLARY 4. A ring has the property that each finitely generated submodule of each of its free modules is free if and only if each finitely generated left ideal is free.

COROLLARY 5. A commutative ring has the property that each finitely generated submodule of each of its free modules is free if and only if it is an integral domain in which each finitely generated ideal is cyclic.

You are now prepared for the main result of this chapter—the theorem that justifies the introduction of semihereditary rings.

THEOREM 7. Each countably generated projective module over a semihereditary ring R is isomorphic to a direct sum of finitely generated left ideals of R.

Proof: By the lemma of §2 and by Corollary 1, to prove this theorem it is enough to show that each element of each projective R-module is contained in a finitely generated direct summand. Thus, we will assume that A is an arbitrary projective R-module and a is an arbitrary element of A. It is remarkably easy to produce the required finitely generated summand of A, containing a, once A has been embedded as a summand in a free R-module: since A is projective, there is a free module F for which $F = A \oplus B$. Relative to any basis of F, a has only a finite number of nonzero coefficients; hence, there is a decomposition $F = F_1 \oplus F_2$, where F_1 and F_2 are free and F_1 is finitely generated and contains a. We have $a \in F_1 \cap A$, and I will show that $F_1 \cap A$ is a finitely generated direct summand of F and consequently also of A. Let p be the projection of F_1 into B with respect to the decomposition $F = A \oplus B$. The kernel of p is $F_1 \cap A$ and the image of p is a finitely generated submodule of the projective module B. By Corollary 3, $\text{im}(p)$ is projective, and consequently $F_1 = (F_1 \cap A) \oplus C$ for some submodule $C \cong \text{im}(p)$. Then $F_1 \cap A$ is finitely generated since it is a homomorphic image (summand) of F_1, and $F = F_1 \oplus F_2 = (F_1 \cap A) \oplus C \oplus F_2$ shows that $F_1 \cap A$ is a summand of F and therefore also of A as required.

Perhaps the most transparent examples of semihereditary rings that are not hereditary are found among the rings of sets that constitute the subject of the next section.

§4 RINGS OF SETS

Let S be an arbitrary set, and let $P(S)$ be the power set of S (that is, the set of all subsets of S). It is easy to verify that $P(S)$ becomes a ring if we define $A + B = (A \cup B)\setminus(A \cap B)$ and $A \cdot B = A \cap B$ for arbitrary subsets A and B of S. The identity element of $P(S)$ is the set S. Each ideal L of $P(S)$ is closed not only under addition and multiplication but also under union since $A \cup B = A + B + A \cdot B$. The objective for this section is to show that $P(S)$ is always semihereditary but not always hereditary. That *$P(S)$ is always semihereditary* is a consequence of the following observation, which contains extra information.

OBSERVATION 3. Each finitely generated ideal L of $P(S)$ is a cyclic direct summand of $P(S)$.

Verification: For an arbitrary subset A of S the ideal generated by A is the set $P(A)$ of all subsets of A and, since $P(S) = P(A) \oplus P(S \setminus A)$, every cyclic ideal is a summand. Now let L be an ideal generated by a finite number of subsets $A_1, \ldots, A_n$ of S. Then L must contain $A = A_1 \cup \cdots \cup A_n$, and, since the ideal generated by A contains each A_i, we have $L = P(A)$. Thus, the finitely generated ideals of $P(S)$ are precisely the cyclic summands of $P(S)$.

From this observation and Theorem 7 we can arrive at a very clear description of the countably generated projective modules over a ring of the form $P(S)$. *Each countably generated projective $P(S)$-module is semiprincipal, and each principal $P(S)$-module is of the form $P(A)$ for some subset A of S.*

If S is finite, then $P(S)$ is semisimple and therefore certainly hereditary (see Exercise 2). *If S is infinite, then $P(S)$ is not hereditary* (see Exercise 7 of Chapter 13). The countably generated ideals of $P(S)$ are always projective (see Exercise 5b), which makes it difficult to demonstrate at this point that $P(S)$ is not hereditary when S is not finite. In § 1 of Chapter 13, we will prove the extended version of Theorem 7: *every projective module over a semihereditary ring is isomorphic to the direct sum of finitely generated left ideals.* If we accept this result, we have the following fact: *every projective $P(S)$-module is semiprincipal.* I will use this fact in the proof of Observation 4, which contains the information that $P(S)$ is not hereditary when S is uncountable.

OBSERVATION 4. If S is uncountable, then the ideal L of $P(S)$, which consists of all the countable subsets of S, is not projective.

Verification: Suppose that L is projective. Then L must be semicyclic, and there is a family $\{A_i \mid i \in I\}$ of countable subsets of S for which $L = \oplus\{P(A_i) \mid i \in I\}$. By the directness of the decomposition, distinct pairs chosen from $\{A_i \mid i \in I\}$ are disjoint. Since each singleton subset of S is in L, $\cup\{A_i \mid i \in I\} = S$. Thus, $\{A_i \mid i \in I\}$ is a partition of S into countable subsets. The index set I must be uncountable. Now let I' be a countably infinite subset of I, and let $A = \cup\{A_i \mid i \in I'\}$. Since A is countable, it is in L. By the construction of A, A cannot be expressed as the union of a finite number of subsets from the various A_i. Thus, the direct decomposition of L is contradicted, and we conclude that L cannot be projective.

The rings of the form $P(S)$ constitute a special subclass of the Boolean rings.

DEFINITION. A ring is *Boolean* if it satisfies the identity $r^2 = r$.

All Boolean rings are semihereditary (see Exercise 6a). Exercises 8 and 9 culminate in a theorem of M. Stone [1936]: *each Boolean ring is isomorphic to a subring of a ring $P(S)$ for some S.* The Boolean rings are a subclass of the class of regular rings.

DEFINITION. A ring R is *regular* if for each r in R there is an r' in R satisfying $rr'r = r$.

For Boolean rings we may use r itself for the required r'. Regular rings are also semihereditary (see Exercise 6d).

NOTES FOR CHAPTER 11

The lemma of § 2 is due to Irving Kaplansky [1958]. Theorem 6 appears in Cartan and Eilenberg [1956]. Theorem 7 (and its extended form, which appears on page 133) belongs to Felix Albrecht [1961]. As a guide to further results related to Theorem 7, see page 41 of Cohn [1971].

EXERCISES

1. Let R be a ring.
   a. Let A_1 and A_2 be R-modules each having the property that all their submodules are countably generated. Show that $A_1 \oplus A_2$ also has this property.
   b. Let S be a submodule of a direct sum $\oplus\{A_i \mid i$ a positive integer$\}$. Show that $S = \cup\{S \cap (A_1 \oplus \cdots \oplus A_k) \mid k$ is a positive integer$\}$.
   c. Show that if each left ideal of R is countably generated then each submodule of each free R-module with a countable basis is countably generated.
   d. Prove Proposition 8.

2. Let S be a finite nonempty set consisting of n elements. Show that $P(S)$ is isomorphic to the ring direct product (as discussed in Chapter 3, § 3) of n copies of the ring Z_2 of integers modulo 2.

3. Let S be a nonempty set.
   a. Show that distinct cyclic (left) ideals of $P(S)$ cannot be isomorphic as $P(S)$-modules.
   b. With each finite (indexed) nest $A_1 \subseteq A_2 \subseteq \cdots \subseteq A_n$ of subsets of S, associate the following $P(S)$-module: $P(A_1) \oplus P(A_2) \oplus \cdots \oplus P(A_n)$. Show that each finitely generated projective $P(S)$-module is isomorphic with the module associated with precisely one such nest.

4. Let A be an R-module with the property that each of its cyclic submodules is a direct summand.
   a. Show that if $A = B \oplus C$ and $a = b + c$ for elements a in A, b in B, and c in C then $A = B \oplus Rc \oplus D$ for some submodule D of C.
   b. Show that each finitely generated submodule of A is a direct summand of A.
   c. Show that each countably generated submodule is the direct sum of cyclic summands of A.

5. Let R be a ring with the property that each cyclic left ideal is a direct summand.
   a. Show that R is semihereditary.
   b. Show that each countably generated left ideal is semiprincipal.

6. a. Show that each Boolean ring is semihereditary.
   b. Suppose R is a ring that contains elements r and r' for which $rr'r = r$. Show that $r'r$ is an idempotent and that $Rr = Rr'r$.
   c. Show that a ring is regular if and only if each of its cyclic left ideals is a direct summand.
   d. Show that each regular ring is semihereditary.

7. Let D be a division ring, and let V be a D-module (that is, vector space over D). Let R be the ring of endomorphisms (that is, the ring of D linear transformations) of V. Show that R is a regular ring.

8. Let R be a Boolean ring.
   a. Show that $x + x = 0$ for each x in R (and therefore that $-x = x$ for each x in R).
   b. Show that R is commutative.
   c. Show that any Boolean field is isomorphic with the field Z_2 of integers mod 2.
   d. Show that if M is any maximal ideal of R then the ring R/M is isomorphic with Z_2.

9. Let R be a Boolean ring.
   a. Show that for any nonzero element a in R, a is not in the ideal $R(1 - a)$.
   b. Let L be the set of all ideals of R that contain $R(1 - a)$ but do not contain the nonzero element a. Let $\mathcal{M}$ be a maximal nest in L, and let $M = \cup\mathcal{M}$. Show that M is a maximal ideal of R that does not contain a.

c. Observe that each nonzero element of R lies outside at least one maximal ideal of R.

d. Let a and b be distinct elements of R, and let M be a maximal ideal of R which does not contain $a - b$ (equal to $a + b$). Use Exercise 8d to show that either a is in M and b is not or b is in M and a is not.

e. Let S be the set of all maximal ideals of R. Let f be the function from R into $P(S)$ defined by $f(a) = \{M \in S \mid a \notin M\}$. Show that f defines an isomorphism of the ring R with the image of f in $P(S)$.

10. Show that if a ring is both semihereditary and quasi-Frobenius then it is semisimple.

11. Show that if a ring is both regular and Noetherian then it is semisimple.

12. Let R be an arbitrary ring. Let M be an R-module that is hereditary (see Exercise 5, page 64) and that satisfies $M \in \text{Inj}(M)$ (see page 72). Show that the ring of endomorphisms of M is regular.

CHAPTER 12

DIRECT SUMS OF COUNTABLY GENERATED MODULES

§ 1 INTRODUCTION

In his study of the structure of projective modules, Kaplansky [1958] found that every projective module must be the direct sum of countably generated modules. He obtained this fundamental result on projectives as a corollary of a theorem that is the subject of the present chapter: *the class of direct sums of countably generated modules over an abitrary ring is closed under direct summands.* The present chapter is an exposition of Kaplansky's proof of this closure property, and the next chapter consists of structural consequences for projective modules that are derived through the use of this closure property.

If we wish, we may regard Kaplansky's closure theorem as a structure theorem in its own right—that is, as a structure theorem for *summands* of direct sums of countably generated modules. However, since we will not develop any further structural insight into arbitrary direct sums of countably generated modules, the fundamental role played by the closure theorem in our study is that of a very powerful lemma for the investigation of projective modules. In future studies of module structure, direct sums of countably generated modules may play a more direct role. In the special case of Z-modules (that is, Abelian groups) this class has already become a subject (and tool) of investigation, as may be seen in Irwin and Richman [1965] and in § 78 of Fuchs [1973].

§ 2 (Ord.) KAPLANSKY'S THEOREM

The purpose of this section is to prove that a direct summand of a direct sum of countably generated modules is again a direct sum of countably generated modules. We begin with the following setting. Let R be an arbitrary ring, and let M be an R-module for which we have two decompositions, $M = A \oplus B$ and $M = \oplus\{M_i \mid i \in I\}$ where I is an arbitrary index set and each M_i is countably generated. We ask what the second decomposition can tell us about A. The approach we take to this question will lead us to consider nests of subsets of I. We can describe the construction of these nests most clearly after introducing (purely for expository purposes) the concept of an *ordinal embedding*. The next paragraph is a description of this concept. The set of all subsets of I is denoted $P(I)$.

I will reserve Greek letters to represent ordinal numbers, and I will use the fact that an ordinal number may be identified with the set of ordinal numbers that strictly precede it. *By an embedding of an ordinal σ in $P(I)$, I mean a function $s:\sigma \to P(I)$ that is one–one and preserves all the unions (or least upper bounds) that exist in σ.* The requirement that s preserve these unions can be interpreted very easily by describing the effect of this requirement on the nature of the image, im(s), of s. The indexed family of subsets $\operatorname{im}(s) = \{s(\alpha) \mid 0 \leq \alpha < \sigma\}$ is a nest for which $s(\alpha)$ is a proper subset of $s(\beta)$ whenever $\alpha < \beta$ and for which $s(\beta) = \cup\{s(\alpha) \mid 0 \leq \alpha < \beta\}$ whenever β is a limit ordinal preceding σ. Each ordinal embedding, s, determines a nest $M_\beta = \oplus\{M_i \mid i \in s(\beta)\}$ $(0 \leq \beta < \sigma)$ of summands of M that will be fundamental in our discussion. (To prevent ambiguity in the meaning of M_β, we assume that the index set I contains no ordinal numbers.) We will be concerned not with arbitrary ordinal embeddings but only with those that are appropriate for the work to be done with respect to the specified decomposition of M. We say that $s:\sigma \to P(I)$ is an *appropriate* (*ordinal*) *embedding* if three conditions hold: (1) $\sigma \geq 1$ and $s(0)$ is the empty set; (2) $M_\alpha = A \cap M_\alpha \oplus B \cap M_\alpha$ $(0 \leq \alpha < \sigma)$; (3) $s(\alpha + 1) \setminus s(\alpha)$ is countable $(0 \leq \alpha < \alpha + 1 < \sigma)$.

Notice that when we work with appropriate ordinal embeddings we have numerous summand relationships. For $0 \leq \alpha < \sigma$, $A \cap M_\alpha$ is a summand of M_α and therefore also of M and A. If $\alpha + 1 < \sigma$, then $A \cap M_\alpha$ is also a summand of $A \cap M_{\alpha+1}$. This multiplicity of summand relationships is at the heart of the "appropriateness" of our specially selected ordinal embeddings. Specifically, we make the following observation:

OBSERVATION 1. If $s : \sigma + 1 \to P(I)$ is an appropriate ordinal embedding, then $A \cap M_\sigma = \oplus\{S_\alpha \mid 0 \leq \alpha < \sigma\}$, where each S_α $(0 \leq \alpha < \sigma)$ is chosen such that $A \cap M_{\alpha+1} = (A \cap M_\alpha) \oplus S_\alpha$.

Verification: The sum $\oplus S_\alpha$ is direct since any violation of directness would violate the directness of $A \cap M_{\alpha+1} = A \cap M_\alpha \oplus S_\alpha$ for some α $(0 \leq \alpha < \sigma)$. The inclusion $A \cap M_\sigma \supseteq \oplus S_\alpha$ is clear from the choices of the S_α. Since $M_0 = 0$, we have $A \cap M_0 \subseteq \oplus S_\alpha$. Suppose $A \cap M_\beta \subseteq \oplus S_\alpha$ for all $\beta < \gamma \leq \sigma$. If γ is a limit ordinal, then $A \cap M_\gamma = A \cap (\cup\{M_\beta \mid \beta < \gamma\})$ $= \cup\{A \cap M_\beta \mid \beta < \gamma\} \subseteq \oplus S_\alpha$, and if γ is not a limit ordinal then $\gamma - 1$ is defined and $A \cap M_\gamma = A \cap M_{\gamma-1} \oplus S_{\gamma-1} \subseteq \oplus S_\alpha$. Thus (by transfinite induction), $A \cap M_\sigma \subseteq \oplus S_\alpha$, which completes the verification.

We have some valuable knowledge about the summands S_α above.

OBSERVATION 2. Referring to the summands S_α $(0 \leq \alpha < \sigma)$ in Observation 1, (1) each S_α is isomorphic to a summand of $\oplus\{M_i \mid i \in s(\alpha + 1) \setminus s(\alpha)\}$, and (2) S_α is countably generated.

Verification: Assertion (1) follows from the following sequence of isomorphisms: $\oplus\{M_i \mid i \in s(\alpha + 1) \setminus s(\alpha)\} \cong M_{\alpha+1}/M_\alpha \cong (A \cap M_{\alpha+1} \oplus B \cap M_{\alpha+1})/(A \cap M_\alpha \oplus B \cap M_\alpha) \cong (A \cap M_{\alpha+1}/A \cap M_\alpha) \oplus (B \cap M_{\alpha+1}/B \cap M_\alpha) \cong S_\alpha \oplus (B \cap M_{\alpha+1}/B \cap M_\alpha)$. Assertion (2) follows from assertion (1), the countability of $s(\alpha + 1) \setminus s(\alpha)$, and the countable generation of the M_i.

We have observed that an ordinal embedding of the form $s : \sigma + 1 \to P(I)$ provides a direct decomposition of the associated submodule $A \cap M_\sigma$ of A. The next natural step would be to try to arrange for $A \cap M_\sigma$ to be as big as possible. We will say that an appropriate ordinal embedding $s : \sigma \to P(I)$ is *maximal* if its graph is not properly contained in the graph of any other appropriate ordinal embedding.

OBSERVATION 3. If $s : \tau \to P(I)$ is a maximal appropriate embedding, then τ is not a limit ordinal; that is, $\tau = \sigma + 1$ for an ordinal number σ.

Verification: If τ were a limit ordinal, we could enlarge s to a function $s' : \tau + 1 \to P(I)$ by defining $s'(\tau) = \cup\{s(\alpha) \mid 0 \leq \alpha < \tau\}$. A thoughtful rereading of the defining conditions of an appropriate embedding will show that s' would be an appropriate embedding properly enlarging s.

OBSERVATION 4. There exists a maximal appropriate embedding $s : \tau \to P(I)$.

Verification: There is always at least one appropriate ordinal embedding—namely, the function $t : 1 \to P(I)$ defined by $t(0)$ equal to the empty set. Let $\mathscr{F}$ be the family consisting of the graphs of the appropriate ordinal embeddings. Let $\mathscr{M}$ be a maximal nest in $\mathscr{F}$, and let G be the union of $\mathscr{M}$. If G is the graph of an appropriate ordinal embedding s, then s must be maximal due to the maximality of $\mathscr{M}$. The set of front coordinates of G is a set of ordinals and is in fact an initial segment of ordinals. Thus, letting τ be the least ordinal that strictly exceeds the ordinals of this initial segment, G is the graph of a one–one function $s : \tau \to P(I)$. We need only show that $s : \tau \to P(I)$ is an appropriate embedding.

Case 1. Suppose τ is not a limit ordinal. Then $\tau = \sigma + 1$ for some σ and σ must be in the domain of an appropriate embedding $s' : \sigma' \to P(I)$ having its graph in $\mathscr{M}$. From the choice of τ and the definition of s it follows that $\sigma' = \tau$ and $s' = s$. In particular, s is an appropriate embedding.

Case 2. Suppose that τ is a limit ordinal. A contradiction will follow by Observation 3 if we show that s is an appropriate embedding. Let α be any ordinal preceding τ. From the construction of τ and the hypothesis that τ is a limit ordinal it follows that there is an ordinal τ_α for which $\alpha + 1 < \tau_\alpha < \tau$ and an $s_\alpha : \tau_\alpha \to P(I)$ having its graph in $\mathscr{M}$. The conditions required for s to be an appropriate embedding are satisfied because they are satisfied by each s_α $(0 \leq \alpha < \tau)$.

The preceding four observations will suggest how to approach the problem of proving the following theorem.

THEOREM 8. For any ring R, each direct summand of each direct sum of countably generated R-modules is itself a direct sum of countably generated submodules.

Proof: Let M be an R-module having decompositions $M = A \oplus B$ and $M = \oplus\{M_i \mid i \in I\}$, where each M_i is countably generated. By Observations 3 and 4, there is a maximal appropriate ordinal embedding $s : \sigma + 1 \to P(I)$. If we can show that $s(\sigma) = I$, then by Observation 1 we will have $A = A \cap M_\sigma = \oplus\{S_\alpha \mid 0 \leq \alpha < \sigma\}$, and we will know that each S_α is countably generated by Observation 2. Thus, we will prove the theorem as follows: *We will assume that $s(\sigma)$ is a proper subset of I, and contradict the maximality of s by constructing an appropriate ordinal embedding $t : \sigma + 2 \to P(I)$ that enlarges s.*

We will begin by locating the single source of difficulty that arises in the construction of an appropriate ordinal embedding $t : \sigma + 2 \to P(I)$ that enlarges s. We must choose a nonempty countable subset J of I, not contained in im(s), in such a manner that $t(\sigma + 1) = s(\sigma) \cup J$, together with $t(\alpha) = s(\alpha)$ $(0 \leq \alpha \leq \sigma)$, will define an appropriate ordinal embedding. A review of the definitions of ordinal embedding and of appropriateness will show that *any* such J will be satisfactory to meet all requirements save one, and that one is $M_{\sigma+1} = A \cap M_{\sigma+1} \oplus B \cap M_{\sigma+1}$. The procedure will be to choose an arbitrary $j \in I \setminus s(\sigma)$ and then to build J by adjoining additional elements of I to $\{j\}$ as needed to insure $M_{\sigma+1} = A \cap M_{\sigma+1} \oplus B \cap M_{\sigma+1}$. Fortunately it will turn out that only a countable subset of I needs to be adjoined to $\{j\}$ to produce this decomposition, and thus the theorem will follow.

Choose an arbitrary j from $I \setminus s(\sigma)$. We will build an infinite matrix consisting of elements of M by choosing the rows in succession.

$$\begin{matrix} x_{1,1} & x_{1,2} & x_{1,3} & x_{1,4} & x_{1,5} & \cdots \\ x_{2,1} & x_{2,2} & x_{2,3} & x_{2,4} & x_{2,5} & \cdots \\ x_{3,1} & x_{3,2} & x_{3,3} & x_{3,4} & x_{3,5} & \cdots \\ x_{4,1} & x_{4,2} & x_{4,3} & x_{4,4} & x_{4,5} & \cdots \\ x_{5,1} & x_{5,2} & x_{5,3} & x_{5,4} & x_{5,5} & \cdots \\ \cdots & \cdots & \cdots & \cdots & \cdots & \end{matrix}$$

The first row is chosen to be a sequence that generates M_j.

The second row is chosen as follows. Let $x_{1,1} = a + b$ with $a \in A$, $b \in B$. With respect to the decomposition $M = \oplus M_i$, $a = \Sigma a_i$ and $b = \Sigma b_i$, where $a_i = b_i = 0$ for all but finitely many i in I. The set $J_{1,1} = \{i \in I \mid a_i \neq 0 \text{ or } b_i \neq 0\}$ is finite. The second row is chosen to be a sequence that generates $\oplus\{M_i \mid i \in J_{1,1}\}$.

The third row is chosen as follows. Let $x_{1,2} = a + b$, $a = \Sigma a_i$, $b = \Sigma b_i$, where $a \in A$, $b \in B$, $a_i \in M_i$, and $b_i \in M_i$. Let $J_{1,2} = \{i \in I \mid a_i \neq 0$

or $b_i \neq 0\}$. The third row is chosen to be a sequence that generates $\oplus\{M_i \mid i \in J_{1,2}\}$.

In choosing the fourth row, we proceed in a similar manner. We do not, however, work with $x_{1,3}$ but with $x_{2,1}$. The pattern will be to proceed down diagonals. At this step we produce a finite subset $J_{2,1}$ of I.

The remaining rows are chosen in the same manner. The succession of elements treated is $x_{1,1}, x_{1,2}, x_{2,1}, x_{1,3}, x_{2,2}, x_{3,1}, \ldots$, and the corresponding finite subsets of I produced are $J_{1,1}, J_{1,2}, J_{2,1}, J_{1,3}, J_{2,2}, J_{3,1}, \ldots$.

We are now ready to define J and complete the proof. Choose $J = \{j\} \cup (\cup\{J_{i,j} \mid i, j \text{ positive integers}\})$. With this choice made the ordinal embedding $t: \sigma + 2 \to P(I)$ is specified (using $t(\sigma + 1) = s(\sigma) \cup J$), and we proceed to verify $M_{\sigma+1} = A \cap M_{\sigma+1} \oplus B \cap M_{\sigma+1}$. Our basic tool will be the pattern of construction of our infinite matrix and the fact that the elements of the matrix generate $\oplus\{M_i \mid i \in J\}$. Since $M_{\sigma+1} = M_\sigma + (\oplus\{M_i \mid i \in J\})$ and $M_\sigma = A \cap M_\sigma \oplus B \cap M_\sigma$, we need only verify that for each x in $\oplus\{M_i \mid i \in J\}$ we have $a, b \in M_{\sigma+1}$, where $x = a + b$, $a \in A$, $b \in B$. To make this last verification, it is enough to consider the special case in which x is one of the elements of our generating set (matrix) $\{x_{i,j} \mid i, j \text{ positive integers}\}$ of $\oplus\{M_i \mid i \in J\}$. Thus, we consider $x_{i,j} = a + b$, $a \in A$, $b \in B$ and ask if a and b are in $M_{\sigma+1}$ as required. Because of the use of the diagonal pattern in the construction of the matrix at some nth step ($n \leq [i + j - 1]^2$) the decomposition $x_{i,j} = a + b$ is taken up, and the submodule generated by the row chosen at this step contains both a and b. In fact, $a, b \in \oplus\{M_k \mid k \in J_{i,j}\} \subseteq M_{\sigma+1}$ as required. This completes the proof of the theorem.

Some interesting generalizations of Kaplansky's theorem have been proved and applied in the study of algebraic questions that are outside the subject of this book. Carol Walker [1966] has given the following generalization. *For an arbitrary ring R and an arbitrary infinite cardinal κ, a summand of a direct sum of R-modules, each of which has a generating set of cardinal not greater than κ, is itself a direct sum of submodules, each of which has a generating set of cardinal not greater than κ.* A generalization that moves into the setting of non-Abelian groups has been given by Paul Hill [1970].

CHAPTER 13

PROJECTIVE MODULES

§ 1 INTRODUCTION

The two previous chapters have left us in a pleasant position. We can draw some substantial conclusions about the structure of projective modules with virtually no effort. Since a projective module is a direct summand of a free module and a free module is certainly a direct sum of countably generated modules (even cyclic modules), Theorem 8 yields the following theorem.

THEOREM 9. For any ring R, every projective R-module is the direct sum of countably generated modules.

By this theorem, the problem of determining the structure of the projective modules over a ring R is reduced to the special case of countably generated projectives. For some rings, the lemma developed in Chapter 11 will then allow this problem to be reduced to the finitely generated case. Specifically, Theorem 9 allows the immediate improvement of Theorem 7.

THEOREM 7 (*completed*). Every projective module over a semihereditary ring is isomorphic to a direct sum of finitely generated left ideals.

Does it seem likely that a theorem on the structure of modules over semihereditary rings would give us new insight into the internal features of hereditary rings? The following corollaries are immediate consequences of Theorem 7; other consequences are indicated in the exercises.

COROLLARY 1. Each left ideal of a hereditary ring is the direct sum of finitely generated left ideals.

COROLLARY 2. A hereditary integral domain is Noetherian.

The remainder of this chapter is devoted to determining the structure of the projective modules over a class of rings that are closely related to division rings.

§ 2 LOCAL RINGS

In a division ring the zero element is the only element that is not a unit—that is, does not have a two-sided multiplicative inverse. Thus, in a very trivial way a division ring has the property that its set of nonunits is a two-sided ideal. A number of other rings also have this property. For each prime integer p and each positive integer n, the nonunits of the ring Z/Zp^n constitute the principal ideal generated by the coset $p + Zp^n$. For the ring consisting of those rational numbers that are expressible with odd denominators, the nonunits constitute the principal ideal generated by 2. Additional examples of rings having this property are included in the exercises. These examples motivate the following definition.

DEFINITION. A ring is *local* if its nonunits form a two-sided ideal.

Let R be a local ring, and let M be the ideal of all nonunits of R. Any left (or right) ideal of R that is proper must consist of nonunits.

Thus, M is the unique maximal left ideal of R, the unique maximal right ideal of R, and the unique maximal two-sided ideal of R. Since M is the *unique* maximal left ideal of R, R has, up to isomorphism, only one simple module—namely, R/M. Since M is a two-sided ideal, R/M is a ring. From the simplicity of the *module* R/M, it follows that the *ring* R/M is a division ring. By Observation 6 of Chapter 2, any two bases of the same (free) R/M-module must have the same cardinal number. Observation 7 of Chapter 2 now applies to give one of the pleasant features of local rings: *any two bases of a free module over a local ring must have the same cardinal number*.

We may think of the local ring R as a close relative of this division ring R/M. For an arbitrary R-module A, it may be appropriate to regard A as a similarly close relative of the associated R/M-space A/MA. That this intuition is appropriate for projective modules is born out by the following theorem of Kaplansky.

THEOREM 10. Each projective module over a local ring is free.

Proof: By Theorem 9 we may restrict our attention to countably generated projectives. The lemma of Chapter 11 and a brief re-examination of the direct sum decomposition made in Observation 2 of that chapter will indicate that to prove the theorem it is sufficient to demonstrate that *each element x of each countably generated projective module P over a local ring R is contained in a finitely generated free summand of P*. Since P is a countably generated projective, it is a direct summand of a countably generated free R-module $F = P \oplus Q$. The procedure for finding a free summand F' of P that contains x will be to construct a basis $v_1, v_2, \ldots, v_i, \ldots$ (i a positive integer) of F for which, for some n, $x = r_1 v_1 + \cdots + r_n v_n$ and $v_1, \ldots, v_n$ are in P. Once this is done we can choose $F' = Rv_1 \oplus \cdots \oplus Rv_n$, and F' will be a summand of P because it is a summand of F. The required basis $\{v_i\}$ will be constructed by modifying a basis $\{u_i\}$ of F, which we will choose with particular attention to x.

Let $\{u_i\}$ be a basis of F with respect to which the representation $x = \Sigma r_i u_i$ of x has the smallest number of nonzero coefficients. Let the indexing of the basis be made so that $x = r_1 u_1 + \cdots + r_n u_n$ with $r_1, \ldots, r_n$ not zero. The desired basis $\{v_i\}$ is constructed by altering only the first n elements of $\{u_i\}$. For each i ($i = 1, \ldots, n$) decompose u_i according to $F = P \oplus Q$ to obtain $u_i = v_i + w_i$ ($1 \leq i \leq n$). Regarding x we have $x = r_1 u_1 + \cdots + r_n u_n = (r_1 v_1 + \cdots + r_n v_n) + (r_1 w_1 + \cdots + r_n w_n)$; since $x, v_1, \ldots, v_n \in P$ and $w_1, \ldots, w_n \in Q$, we have simply $x = r_1 v_1 + \cdots + r_n v_n$. We let F' be the submodule generated by $v_1, \ldots, v_n$. We have $x \in F' \subseteq P$, and to complete the proof it will be sufficient to verify that $v_1, \ldots, v_n, u_{n+1}, u_{n+2}, \ldots$ is a basis of F.

First we must lay out an important tool. The choice of the indexed basis $\{u_i\}$ gives us a condition on the coefficients r_i : *no r_i is a right linear combination of the remaining coefficients.* To justify this assertion it is enough to consider the notationally most convenient case, r_n. Suppose $r_n = r_1 r'_1 + \cdots + r_{n-1} r'_{n-1}$. Then $x = r_1 u_1 + \cdots + r_{n-1} u_{n-1} + (r_1 r'_1 + \cdots + r_{n-1} r'_{n-1}) u_n = r_1(u_1 + r'_1 u_n) + \cdots + r_{n-1}(u_{n-1} + r'_{n-1} u_n)$ is an expression for x involving only $n - 1$ nonzero coefficients. Since the sequence $u_1 + r'_1 u_n, \ldots, u_{n-1} + r'_{n-1} u_n, u_n, u_{n+1}, \ldots$ can be mentally verified to be a basis of F, we have contradicted the choice of $\{u_i\}$. Thus, we have justified the italicized assertion.

For each v_i $(1 \leq i \leq n)$ we have a unique representation

$$v_i = r_{i1} u_1 + \cdots + r_{in} u_n + t_i$$

where t_i is a linear combination of u_j for which $j > n$. We need some information about the coefficients r_{ij}. To get it, we replace each v_i in the equation $r_1 v_1 + \cdots + r_n v_n = r_1 u_1 + \cdots + r_n u_n$ by its expansion in terms of the basic $\{u_i\}$. Equating coefficients of each u_i then gives n equations: $r_1 r_{1i} + r_2 r_{2i} + \cdots + r_n r_{ni} = r_i$ $(1 \leq i \leq n)$. Rearranging gives $r_1 \cdot r_{1i} + \cdots + r_{i-1} \cdot r_{(i-1)i} + r_i \cdot (r_{ii} - 1) + r_{i+1} \cdot r_{(i+1)i} + \cdots + r_n \cdot r_{ni} = 0$. Since no r_j is a right linear combination of the others, we conclude that the following elements of R are nonunits: $r_{1i}, \ldots, r_{(i-1)i}, (r_{ii} - 1), r_{(i+1)i}, \ldots, r_{ni}$. Since we have this conclusion for each i $(1 \leq i \leq n)$, we may summarize that r_{ij} is a nonunit if $i \neq j$ and $1 - r_{ii}$ is a nonunit for each i. Thus far our discussion has been valid for arbitrary rings. We now make our first use of the fact that R is local: since $1 = r_{ii} + (1 - r_{ii})$ and $1 - r_{ii}$ is a nonunit, we conclude that r_{ii} is a unit.

To complete the proof of the theorem it is sufficient to show that *if $u_1, u_2, \ldots, u_i, \ldots$ (i a positive integer) is a basis of a free module F over a local ring R and*

$$(1) \qquad \begin{aligned} v_1 &= r_{11} u_1 + \cdots + r_{1n} u_n + t_1 \\ &\vdots \\ v_n &= r_{n1} u_1 + \cdots + r_{nn} u_n + t_n \end{aligned}$$

is a system of equations for which r_{ij} is a unit if and only if $i = j$ and each t_i is a linear combination of the u_j $(j > n)$, then $v_1, \ldots, v_n, u_{n+1}, u_{n+2}, \ldots$ is also a basis of F. This will be proved by induction on n. Assume that the corresponding assertion in which n is replaced by $n - 1$ is true. Since r_{nn} is a unit, it is mentally verifiable that $u_1, \ldots, u_{n-1}, v_n, u_{n+1}, u_{n+2}, \ldots$ is a basis of F. (This also verifies the assertion for $n = 1$.) In the system (1) the last equation can be solved for u_n and the result substituted into the preceding equations to produce a system

$$v_1 = r'_{11}u_1 + \cdots + r'_{1(n-1)}u_{n-1} + s_1$$
$$\vdots \qquad\qquad\qquad\qquad \vdots$$
$$v_{n-1} = r'_{(n-1)1}u_1 + \cdots + r'_{(n-1)(n-1)}u_{n-1} + s_{n-1}$$

where each s_i is a linear combination of v_n and the u_j ($j > n$). In order to employ our induction hypothesis we must show that r'_{ij} is a unit if and only if $i = j$. To compute r'_{ij} let r be the inverse of r_{nn}. Then $r'_{ij} = r_{ij} - r_{in}rr_{nj}$. Since neither i nor j is equal to n, $r_{in}rr_{nj}$ is certainly a nonunit. We use the hypothesis that R is local a second time: since $r_{in}rr_{nj}$ is a nonunit, $r_{ij} - r_{in}rr_{nj}$ is a unit if and only if r_{ij} is a unit. Thus, r'_{ij} is a unit if and only if $i = j$. We can now apply our induction hypothesis and conclude that, since $u_1, \ldots, u_{n-1}, v_n, u_{n+1}, u_{n+2}, \ldots$ is a basis of F, $v_1, \ldots, v_{n-1}, v_n, u_{n+1}, u_{n+2}, \ldots$ is also a basis of F. This completes the proof of the theorem.

A summary of our structure theorems for projective modules is included in § 2 of the next chapter.

NOTES FOR CHAPTER 13

Theorems 9 and 10 are due to Kaplansky [1958] and Theorem 7 is due to Albrecht [1961]. The dependence of Chapters 11, 12, and 13 on Kaplansky's paper can hardly be overemphasized. These three chapters are essentially an embedding of Albrecht's paper into Kaplansky's, with the result being incorporated into the program of the present book.

There are two important classes of rings to which I feel I must relate the study we have made in this book. These two classes of rings are the Dedekind domains and the Artinian rings. Theorem 7 and Corollary 2 of this chapter give me sufficient reason to discuss the Dedekind domains here. I will define and discuss the Artinian rings in the notes to Chapter 14, but I do want to remark here that the projectives over Artinian rings are of known structure: they are all semiprincipal.

The study of Dedekind domains has a long history, and these domains can be defined in several equivalent ways. We will use the following definition: *a Dedekind domain is a hereditary integral domain.* The problem of describing the structure of the projectives and injectives over a Dedekind domain R is reducible to considerations involving the ideals of

R: by Theorem 7 every projective must be isomorphic with a direct sum of (necessarily finitely generated) left ideals of R. Moreover since R is a domain, every left ideal of R must be uniform and consequently the projectives are also semiuniform. By Corollary 2, R is Noetherian, and consequently by Theorem 3 of Chapter 9 each injective R-module is semiuniform. By Exercises 11 through 17 on page 108 (or by the cataloging procedure of § 2, Chapter 9), each uniform injective R-module is isomorphic with the injective hull of R/P for a unique prime ideal P of R.

By consulting Cartan and Eilenberg [1956] or Rotman [1970] you can see how the definition of Dedekind domain can be reformulated in terms of "invertibility" of ideals. From there you can go to Curtis and Reiner [1962] or Zariski and Samuel [1958] to find a theorem to the effect that each ideal of a Dedekind domain is uniquely expressible as a product of prime ideals. This theorem explains the long-standing interest in Dedekind domains. You can also find (see page 279 of Zariski and Samuel [1958]) that each ideal of a Dedekind domain can actually be generated by two or fewer elements.

EXERCISES

1. Let R be a ring. By a *formal power series* in x over R we mean an expression $a_0 + a_1x + a_2x^2 + \cdots$, where the a_i are elements of R. Such expressions will be denoted briefly by $\Sigma a_i x^i$. The set of all formal power series over R is denoted $R[[x]]$. Addition and multiplication for these series are defined by $\Sigma a_i x^i + \Sigma b_i x^i = \Sigma(a_i + b_i)x^i$ and $\Sigma a_i x^i \cdot \Sigma b_i x^i = \Sigma c_i x^i$, where the coefficients c_i are given by $c_i = \Sigma\{a_j b_k \mid j + k = i\}$.
   a. Verify that $R[[x]]$ is a ring.
   b. Show that $a_0 + a_1x + a_2x^2 + \cdots$ is a unit in R if and only if $a_0 \neq 0$.
   c. Show that $R[[x]]$ is local if and only if R is local.
2. Let D be a division ring.
   a. Show that projective $D[[x]]$-modules must be free.
   b. Show that submodules of free $D[[x]]$-modules are necessarily free.
3. Let D be a division ring, and let $D[[x, y]]$ denote the ring $(D[[x]])[[y]]$—that is, the ring of formal power series in y over the ring $D[[x]]$.
   a. Show that projective $D[[x, y]]$-modules must be free.
   b. Show that any two bases of a free $D[[x, y]]$-module must have the same cardinal numbers.
   c. Show that the (left) ideal of $D[[x, y]]$ generated by the two elements x and y is not projective.
4. In Exercise 1 we denoted the elements of $R[[x]]$ by the formal expressions $\Sigma a_i x^i$ with the understanding that i denotes a *nonnegative* integer. Let us now relax this restriction and examine the formal expressions of the form $\Sigma a_i x^i$,

where i is allowed to assume *arbitrary* integer values. Addition of such expressions may be defined as $\Sigma a_i x^i + \Sigma b_i x^i = \Sigma(a_i + b_i)x^i$. The previous definition of multiplication breaks down as follows. In the computations of the coefficient c_i for the product expression, the sum $\Sigma\{a_j + b_k \mid j + k = i\}$ may involve an infinite number of nonzero terms and therefore fail to be defined. Let $R\langle x \rangle$ consist of those $\Sigma a_i x^i$ for which the set $\{i \mid i < 0, a_i \neq 0\}$ is finite. For elements of $R\langle x \rangle$ the expressions for c_i are defined, and consequently we have operations of addition and multiplication in $R\langle x \rangle$ that extend the operations of $R[[x]]$. (Help is available in McCoy [1964].)

a. Verify that $R\langle x \rangle$ is a ring.
b. Show $R\langle x \rangle$ is a division ring if and only if R is a division ring.
c. Let D be a division ring. Since $D\langle x \rangle$ is a ring containing $D[[x]]$ as a subring, $D\langle x \rangle$ may be regarded as a $D[[x]]$-module. Show that the $D[[x]]$-module $D\langle x \rangle$ is an injective (divisible) hull of the $D[[x]]$-module $D[[x]]$.

5. Let R be an arbitrary ring, and let A be a uniform injective R-module. Let E be the ring of endomorphisms of A (as explained in Exercise 12 of Chapter 2). Show that E is a local ring.

6. a. Show that if a ring is both local and regular then it is a division ring.
   b. Show that if a commutative ring is both local and hereditary then it is a principal ideal domain.

7. Let S be a nonempty set.
   a. Show that if A is a proper projective (left) ideal of $P(S)$ and A is not of the form $P(S \setminus \{s\})$ for any s in S then A is properly contained in a proper projective ideal of $P(S)$.
   b. Let F be the set of all finite subsets of S. Verify that F is an ideal of $P(S)$.
   c. Assume now that S is infinite. Show that no maximal ideal of $P(S)$ that contains F can be projective.
   d. Show that $P(S)$ is hereditary if and only if S is finite.

8. A function $f: C \to C$ from the plane C of complex numbers into itself is an *entire function* if f possesses a derivative at each point of C.
   a. Verify that the set E of all entire functions forms a ring with respect to the operations $(f + f')(c) = f(c) + f'(c)$ and $(f \cdot f')(c) = f(c) \cdot f'(c)$.
   b. When are two cyclic (left) ideals of E isomorphic?
   c. Describe the finitely generated ideals of E. (See Helmer [1940] for the solution.)
   d. Show that E is semihereditary.
   e. Describe the projective E-modules.
   f. Show that E is neither hereditary nor regular.

REVIEW

CHAPTER 14

SUMMARY

§ 1 INTRODUCTION

The following chart outlines the classes of modules for which we have developed the most precise structural descriptions. The four horizontal lines that cross the chart and have labels at the extreme right are to be interpreted as follows. The top line is labeled "projectives" and indicates that we have good structural results for the projective modules over each of the types of rings listed on the chart. The second line is labeled "injectives" and indicates that we have good structural results for the injective modules over each of the types of rings listed below this line. The meaning of the two remaining lines is now clear. Our interest will be centered on reviewing the results we have obtained for projectives and injectives.

The true way to review this book is to choose a specific ring R and attempt to describe the projective, injective, finitely generated, and other R-modules by applying results or techniques given in the book.

A CHART OF CLASSES OF MODULES HAVING SATISFACTORILY PRECISE STRUCTURAL DESCRIPTIONS

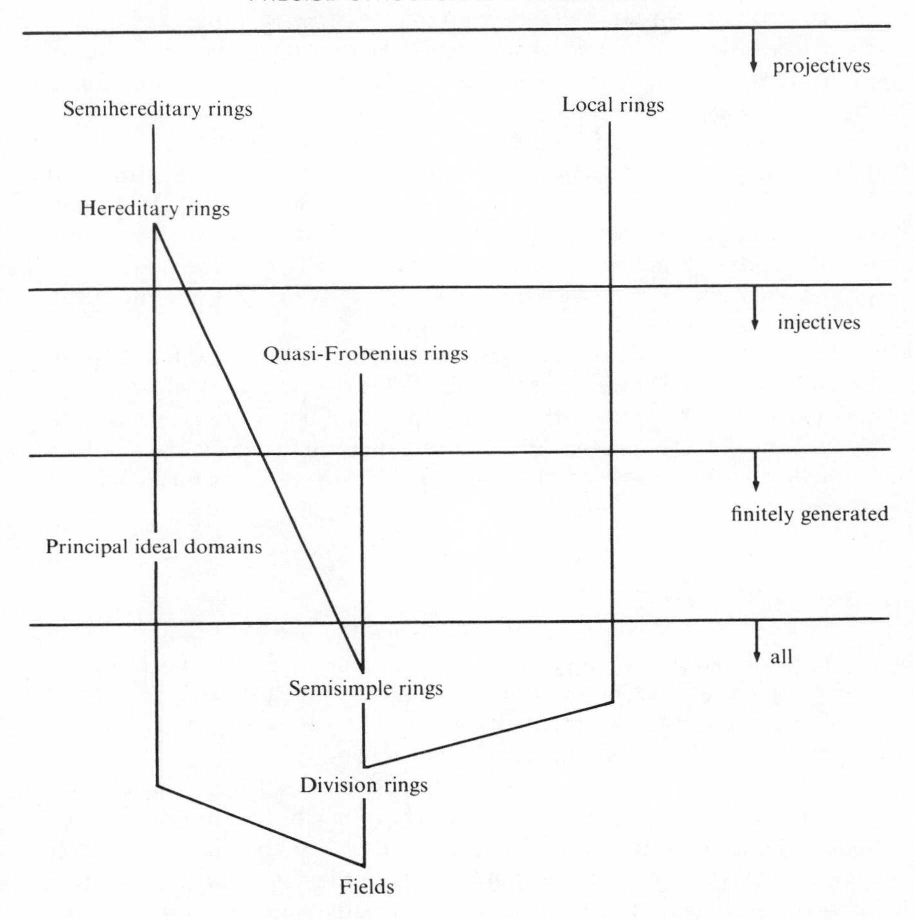

How encouraging these attempts will prove to be will vary greatly with the choice of *R*. In the Example Projects, I have suggested some rings for which several classes of modules admit concise descriptions when our results and methods are used.

§ 2 PROJECTIVES

We have found that, if a ring *R* is either local, quasi-Frobenius, or semihereditary, then each projective *R*-module is the direct sum of finitely generated submodules, each of which is isomorphic to a (projective)

left ideal. If R is either local or a principal ideal domain, then the result is especially strong: each projective module is free. Note that, in either case, R regarded as an R-module is indecomposable. When R is quasi-Frobenius, there is a little more variety: each projective module is the direct sum of uniform principal modules. Note that, in this case, the uniform principal modules are indecomposable cyclics.

When R is hereditary, each submodule of each free R-module is projective, and consequently for these rings we know the structural possibilities for arbitrary submodules of free modules. When R is semi-hereditary, we know that finitely generated submodules of free R-modules are projective. Finally, we recall one very substantial theorem about projective modules over arbitrary rings: every projective module is the direct sum of countably generated modules.

At the risk of overemphasizing the extent to which our structural investigations have been concerned with projectivity, I suggest that you place your hand over the words "principal ideal domains" on the chart. Now observe that for the remaining rings on the chart, the modules belonging to the classes whose structure we have determined are all projective.

§ 3 INJECTIVES

Our structural results for injective modules may be reviewed simply by rereading Chapter 9 in which all these results were presented. I will comment only on the contrast between the descriptions used to specify the structure of projectives and injectives.

The structure theory for injectives over Noetherian rings is in a very satisfying state; for this reason, our chart may be deceptive in that we have not included an entry "Noetherian rings" under the line labeled "injectives." However, there is a difference in the degree of transparency between the structural description of a projective module over a ring of one of the types listed on the chart and an injective module over an arbitrary Noetherian ring. The projectives referred to are built from (finitely generated projective) left ideals, whereas the injectives are built from the *injective hulls* of (uniform) cyclics. Opaqueness may develop in the passage from the relevant cyclics to their injective hulls. In some cases these hulls can be described with perfect clarity. We found this to be the case for modules over principal ideal domains. For quasi-Frobenius rings we were able to avoid the problem of building hulls by deriving the structure of the projective = injective modules directly from the interplay between the properties of projectivity and injectivity.

NOTES FOR CHAPTER 14

Before closing this book I feel that I must discuss the work we have done here in relation to a very major class of rings—the Artinian rings. My remarks will include material that is not covered in this book but that will seem elementary to many readers. We will use the following definition, which should be compared with the characterization of Noetherian rings given in Exercise 2 of Chapter 7: *a ring is Artinian if it contains no infinite nest of distinct left ideals of the form* $L_1 \supset L_2 \supset \cdots \supset L_i \supset \cdots$ (*i a positive integer*). Surprisingly, perhaps, every Artinian ring is necessarily Noetherian (see page 368 of Curtis and Reiner [1962]). It follows that each Artinian ring R has a composition (that is, unrefinable) series of left ideals $R = L_1 \supset \cdots \supset L_i \supset \cdots \supset L_m = 0$. From the Jordan–Holder theorem (see page 83 of Curtis and Reiner [1962] or page 21 of Lambek [1966]), it follows that every simple R-module must be isomorphic to one of the modules $L_1/L_2, \ldots, L_i/L_{i+1}, \ldots, L_{n-1}/L_n$. In particular, an Artinian ring can have only finitely many nonisomorphic simple modules. It also follows from the Jordan–Holder theorem that every nonzero cyclic R-module (hence *every* nonzero R-module) must contain a simple submodule. With these remarks made, it is convenient to discuss the projectives and injectives for Artinian rings.

Let R be an Artinian ring. Then R can be decomposed into a direct sum $R = L_1 \oplus \cdots \oplus L_n$ where each L_i is an indecomposable principal left ideal. The Krull–Schmidt theorem (see page 79 of Curtis and Reiner [1962] or page 24 of Lambek [1966]) then applies to yield the conclusion that each indecomposable principal R-module must be isomorphic to (at least) one of these L_i. In particular we see that an Artinian ring can have only finitely many nonisomorphic principal indecomposable modules. Since R possesses a composition series, the same is true of each finitely generated free R-module. The Krull–Schmidt theorem applies again to yield the conclusion that each finitely generated projective module over an Artinian ring is semiprincipal. It is in fact true (but not proved here) that *every* projective module over an Artinian ring is semiprincipal. For a proof that this conclusion follows for an even broader class of rings, the semiperfect rings, see Eilenberg, Nagao, and Nakayama [1955] or Mueller [1970]. Consider now the injectives. Since Artinian rings are necessarily Noetherian, each injective R-module is semiuniform. Since each nonzero R-module must contain a simple module, we may describe the uniform injective R-modules as "the injective hulls of the simple R-modules." By Observation 3 of Chapter 9 (or directly), the injective hulls of two simple modules will be isomorphic if and only if the simple modules themselves are isomorphic. Thus for an Artinian ring R we have the following classification of uniform injective R-modules: there is a one–one correspondence between the isomorphism types of uniform injective R-modules and the isomorphism types of simple R-modules. Since there are only a finite number of the latter, there are only a finite number of the former.

The situation regarding the projectives and injectives for Artinian rings is somewhat reminiscent of the situation for quasi-Frobenius rings. It is in fact true (but not proved here) that quasi-Frobenius rings are necessarily Artinian.

EXAMPLE PROJECTS

1. Let D be a division ring. Let $D[[x]]$ be the ring of formal power series in x over D as defined in Exercise 1 of Chapter 13. Using the results and techniques of this book, go as far as you can in determining the structure of $D[[x]]$-modules. Your results should include descriptions of the projectives, the submodules of free modules, the injectives, the finitely generated modules, and perfectly precise descriptions of the cyclics and the uniform injectives.

2. Let Z_{p^n} be the ring of integers modulo p^n, where p is a prime and n is a positive integer. Study the class of Z_{p^n}-modules as follows:
   a. Determine the projectives and injectives.
   b. Show that a Z_{p^n}-module that contains no nonzero injective submodule possesses a natural $Z_{p^{n-1}}$-module structure. (See Exercise 13 of Chapter 2.)
   c. Show that for an arbitrary Z_{p^n}-module A, $A = B \oplus C$, where B is injective and C contains no nonzero injective submodule. (See Exercise 4 of Chapter 8.)
   d. Using the results above, prove by induction on n that every Z_{p^n}-module is semicyclic.

   Let Z_n be the ring of integers modulo n, where n is a positive integer. Using Exercise 2d and an argument in the style of Euclid (primary decomposition), show that every Z_n-module is semicyclic.

   Let R be a principal ideal domain, and let I be a nonzero ideal of R. Show that every R/I-module is semicyclic.

3. a. Reread each observation, proposition, theorem, and corollary in this book with the following question in mind: "What can I make this result tell me about Abelian groups?"
   b. Search this book to find as many uses, characterizations, and sources of examples of division rings as possible.

BIBLIOGRAPHY

Albrecht, F. On projective modules over semi-hereditary rings. *Proceedings of the American Mathematical Society*, 12 (1961), 638–639.

Artin, E. The influence of J. H. M. Wedderburn on the development of modern algebra. *Bulletin of the American Mathematical Society*, 56 (1950), 65–72.

Baer, R. Abelian groups that are direct summands of every containing Abelian group. *Bulletin of the American Mathematical Society*, 46 (1940), 800–806.

Bass, H. Finitistic dimension and a homological generalization of semi-primary rings. *Transactions of the American Mathematical Society*, 95 (1960), 466–488.

Bass, H. *Algebraic K-Theory*. New York: Benjamin, 1968.

Boerner, H. *Representations of groups, with special consideration for the needs of modern physics*. Amsterdam: North Holland Publishing Co., 1963.

Burrow, M. *Representation theory of finite groups*. New York: Academic Press, 1965.

Cartan, E., & S. Eilenberg. *Homological algebra*. Princeton, N.J.: Princeton University Press, 1956.

Cohn, P. M. *Free ideal rings and their relations*. New York: Academic Press, 1971.

Cotton, F. O. *Chemical applications of group theory*. 2nd ed. New York: Wiley-Interscience, 1971.

Curtis, C. W., & I. Reiner. *Representation theory of finite groups and associative algebras*. New York: Wiley-Interscience, 1962.

Eckmann, B., & O. Schopf. Über injektive moduln. *Archiv. der Math.*, 4 (1953), 75–78.

Eilenberg, S., H. Nagao, & T. Nakayama. On the dimension of modules and algebras, II. *Nagoya Mathematical Journal*, 9 (1955), 1–16.

Faith C. Rings with ascending condition on annihilators. *Nagoya Mathematical Journal*, 27 (1966), 179–191.

Faith, C. *Rings, modules, and categories, I*. New York: Springer-Verlag, 1972.

Faith, C., & E. Walker. Direct-sum representations of injective modules. *Journal of Algebra*, 5 (1967), 203–221.

Freyd, P. *Abelian categories: an introduction to the theory of functors*. New York: Harper & Row, 1964.

Fuchs, L. *Infinite Abelian groups*. 2 vols. New York: Academic Press, 1970–1973.

Hall, Lowell H. *Group theory and symmetry in chemistry*. New York: McGraw-Hill, 1969.

Halmos, P. R. *Naive set theory*. New York: Van Nostrand, 1960.

Hartley, B., & T. O. Hawkes. *Rings, modules, and linear algebra*. London: Chapman and Hall Ltd., 1970.

Helmer, O. Divisibility properties of integral functions. *Duke Mathematical Journal*, 6 (1940), 345–356.

Herstein, I. *Topics in algebra*. Waltham, Mass.: Blaisdell, 1964.

Hill, P. On the decomposition of certain infinite nilpotent groups. *Mathematische Zeitschrifte*, 113 (1970), 237–248.

Hilton, P. Lectures in homological algebra. *Regional Conference Series in Mathematics, No. 8*. Providence, R.I.: American Mathematical Society, 1971.

Hu, S. T. *Elements of modern algebra*. San Francisco: Holden-Day, 1965.

Hu, S. T. *Introduction to homological algebra*. San Francisco: Holden-Day, 1968.

Irwin, J., & F. Richman, Direct sums of countable groups and related concepts. *Journal of Algebra*, 2 (1965), 443–450.

Jans, J. P. Projective injective modules. *Pacific Journal of Mathematics*, 9 (1959), 1103–1108.

Kalman, R. E. P., P. L. Falb, & M. O. Arbib. *Topics in mathematical system theory*. New York: McGraw-Hill, 1969.

Kaplansky, I. Projective modules. *Annals of Mathematics*, 68 (1958), 372–377.

Kaplansky, I. *Infinite Abelian groups*. 2nd. ed. Ann Arbor: University of Michigan Press, 1969.

Kelley, J. L. *General topology*. New York: Van Nostrand, 1955.

Kimberling, C. H. Emmy Noether. *American Mathematical Monthly*, 79 (1972), 136–149.

Lambek, J. *Lectures on rings and modules*. Waltham, Mass.: Blaisdell, 1966.

MacLane, S. *Categories for the working mathematician*. New York: Springer-Verlag, 1971.

MacLane, S., & G. Birkhoff. *Algebra*. New York: Macmillan, 1967.

Maschke, H. Über den arithmetischen Charakter der Coefficienten der Substitutionen endlicher linearer Substitutionsgruppen. *Math. Ann.*, 50 (1898), 482–498.

Matlis, E. Injective modules over Noetherian rings. *Pacific Journal of Mathematics*, 8 (1958), 511–528.

McCoy, N. *The theory of rings*. New York: Macmillan, 1964.

Mitchell, O. R., & R. W. Mitchell. *An introduction to abstract algebra*. Monterey, Calif.: Brooks/Cole, 1970.

Mueller, B. J. On semi-perfect rings. *Illinois Journal of Mathematics*, 14 (1970), 464–467.

Osofsky, B. A generalization of quasi-Frobenius rings. *Journal of Algebra*, 4 (1966), 373–387.

Richman, F. *Number theory: An introduction to algebra*. Monterey, Calif.: Brooks/Cole, 1971.

Rotman, J. *The theory of groups: An introduction*. Boston: Allyn & Bacon, 1965.

Rotman, J. *Notes on homological algebra*. New York: Van Nostrand Reinhold, 1970.

Rubin, H., & J. Rubin. *Equivalents of the axiom of choice*. Amsterdam: North Holland Publishing Company, 1963.

Rutter, E. Two characterizations of QF rings. *Pacific Journal of Mathematics*, 30 (1969), 777–784.

Rutter, E. PF modules. *Tôhoku Mathematical Journal*, 23 (1971), 201–206.

Rutter, E. A characterization of QF-3 rings. *Pacific Journal of Mathematics* (forthcoming).

Schreier, O., & E. Sperner. *Introduction to modern algebra and matrix theory*. New York: Chelsea, 1959.

Sharpe, D. W., & P. Vamos. Injective modules. *Cambridge Tracts in Mathematics*, No. 62. New York: Cambridge University Press, 1972.

Stone, M. H. The theory of representations of Boolean algebras. *Transactions of the American Mathematical Society*, 40 (1936), 37–111.

Walker, C. P. Relative homological algebra and Abelian groups. *Illinois Journal of Mathematics*, 10 (1966), 186–209.

Zariski, O., & P. Samuel. *Commutative algebra*. 2 vols. New York: Van Nostrand, 1958–1960.

INDEX